Signature of Excellence

Ball Corporation at 125

90
120
JPL
NASA
NASA / GODDARD
SPACE FLIGHT CENTER
Electro-Optics /
Cryogenics
Division

Signature of Excellence

Ball Corporation at 125

RICHARD BLODGETT

GREENWICH PUBLISHING GROUP, INC.
OLD SAYBROOK, CONNECTICUT

Produced and published by Greenwich Publishing Group, Inc.
Old Saybrook, Connecticut
www.greenwichpublishing.com

Designed by Clare Cunningham Graphic Design

Library of Congress Control Number: 2005932945

ISBN: 0-944641-63-6

First Printing: November 2005

10 9 8 7 6 5 4 3 2 1

PHOTO CREDITS:
All images appear courtesy of Ball Corporation, except the following:
Page 33 (lower right) appears courtesy of Northwestern University Library
Page 40 appears courtesy of NASA
Page 46 © *Indiana Business Magazine*
Page 55 Reproduction of front page courtesy of *The Star Press*
Page 109 appears courtesy of NASA/JPL-Caltech/O. Krause (Steward Observatory)
Page 118 (upper) appears courtesy of NASA/JPL-Caltech
Page 119 appears courtesy of NASA/JPL-Caltech/UMD
Page 121 appears courtesy of DigitalGlobe

Table of Contents

HANDS OFF
PLEASANT VALLEY
Mother-Daughter
Jackson

This book celebrates Ball Corporation's 125th anniversary. It is dedicated to employees worldwide, whose ideas, commitment and focus on the customer have always been the reason for our success in both packaging and aerospace.

Ball Corporation: The First Century, published in 1980, told the story of Ball's first one hundred years. This book relates that story but emphasizes the past 25 years, a time during which Ball has changed dramatically. It tells who we are, how we serve our customers and where we are heading. I invite you to read the pages that follow and learn more about our history, our people and our values.

— R. David Hoover

Chairman, president and chief executive officer

Ball

CHAPTER ONE:

"We Want to Be The Best"

There is nothing mysterious or flashy about the success of Ball Corporation. It prospers year after year by turning out billions of recyclable metal cans and plastic bottles efficiently while providing a high level of service to its packaging customers. At the same time, it prospers in a vastly different business, aerospace, by solving complex technical challenges for NASA, defense and commercial organizations. These two operations — packaging and aerospace—are united in their focus on the customer. In combination, they have given Ball an outstanding growth record.

"We constantly do little things better and better and pay close attention to our customers," says Jack Pickenbrock, senior vice president, manufacturing, in Ball's metal beverage container operations. In some circles, this approach may be considered mundane. At Ball, however, the ability to mass-produce food and beverage containers to extremely close tolerances and earn a profit doing so has made it one of the world's great manufacturing companies—a source of admiration in its industry and a point of pride among employees.

The company's Ball Aerospace & Technologies Corporation business is a highly respected developer of customized technology used to explore the deepest reaches of space, monitor weather patterns and provide communications and surveillance. "We build things people have never built before to do things people have never done before," says David L. Taylor, president and CEO of Ball Aerospace.

Ball is one of the world's most successful manufacturers of metal and plastic packaging and a leader in advanced aerospace technology for government and commercial customers. It has prospered through the teamwork of its people, who work closely with customers. "How our people do their jobs every day is what makes us successful," says Chairman, President and CEO Dave Hoover.

Ball has developed new packaging products such as Heat-Tek® PET plastic bottles, below, for juices, teas, sports drinks and other hot-filled beverages. The company is North America's largest manufacturer of aluminum beverage cans—used every day by millions of people, including the young soccer players, opposite.

1980-2005: Critical Period of Change

Former Ball CEO Richard M. "Dick" Ringoen used to say, "It's a heck of an accomplishment to be more than 100 years old and still have your name on the door. The only way you do that is by changing successfully."

As Ball celebrates its 125th anniversary in 2005, this book recounts the company's rich history—paying particular attention to its ability since 1980 to reshape itself in response to changing markets—and looks ahead to the opportunities available to Ball and its people in the company's ever-evolving businesses.

The period since 1980 demonstrates Ball's resilience as well as its long-term success. From 1980 through 1987, the company capitalized on robust markets to achieve consistently higher earnings. Beginning in 1988, its markets weakened and earnings fell. Yet in the ensuing decade the company found new opportunities, made important acquisitions, grew organically, divested non-core businesses and reduced costs. Since 1998 it has emerged stronger and more profitable than ever.

Earnings of $295.6 million in 2004 were nearly double those two years earlier, while sales of $5.44 billion were up more than 40 percent in two years. The *Denver Post* said in 2003 that Ball's success demonstrates "how solid business practices, focus and foresight not only can sustain a corporation in tough times but also make it robust." It praised the company for its "patient and methodical management."

Unique Formula: Packaging and Aerospace

Ball has staked out major positions in markets in which it has the know-how, customer contacts and financial resources to compete effectively. Chairman, President and CEO R. David "Dave" Hoover says, "We don't necessarily want to be the biggest, we want to be the best in each of our markets by serving our customers."

The company—based in Broomfield, Colorado—is the largest manufacturer of aluminum beverage cans in North America and one of the two largest beverage can manufacturers in the world. It supplies cans for Budweiser, Coca-Cola, Coors, Heineken, Miller, Molson, Pepsi-Cola and other well-known brands in North America; Coca-Cola, Pepsi-Cola, Heineken, Stella Artois, Carling and Carlsberg in Europe; Skol, Brahma and Itaipava in Brazil; and Jianlibao Orange & Honey Sports Drink, Tsingtao and Yanjing in the People's Republic of China, among others.

All told, the company manufactures more than 50 billion recyclable beverage cans every year, equal to seven per year for every man, woman and child on Earth. Senior Vice President and Chief Financial Officer Raymond J. Seabrook remarks, "If placed end to end at the equator, the cans we produce each year would go around the world more than 100 times. That's a lot of cans."

While beverage cans are Ball's largest business, accounting for about two-thirds of revenues, Ball is also the second largest manufacturer of steel food cans in North America. Top-selling brands such as Allens Cut Italian Green Beans, Bush's Beans, Chef Boyardee Spaghetti & Meatballs, Red Gold Premium Tomatoes

Ball Aerospace is designing and building the WorldView orbiting camera, above, which will collect high-resolution images of Earth in eight different spectral bands for the U.S. National Geospatial-Intelligence Agency (NGA).

and Trident Foods are sold in Ball-made cans. In addition, Ball is an important producer of plastic bottles for Pepsi-Cola, Vermont Pure purified drinking water, Coors beer and popular new beverages such as energy drinks, ready-to-drink teas and vitamin-enhanced water. In 2004, *Fortune* magazine wrote, "You've probably never heard of Ball Corp., but chances are you use its products all the time."

Ball Aerospace's products, by contrast, are used in the scientific and defense communities. Ball Aerospace is known for its leadership in remote-sensing technologies and has participated in some of the most complex ventures of our time—such as providing instruments for new satellite-based surveillance and weather-imaging systems, supplying antenna assemblies and electronics for the Mars Exploration Rovers and building the pair of spacecraft for Deep Impact, one of which successfully crashed into Comet Tempel 1 in 2005.

No other company in the world combines major positions in packaging and aerospace—from the earthly (cans and bottles) to the celestial (spacecraft). Both are comfortably at home within Ball Corporation. Ball's consistently strong performance in two very different businesses shows how skilled, motivated people can transform an old-line company into a dynamic contemporary business.

Ball's Personality and Values

Companies, like people, have distinct personalities. Four themes that recur in this book help define the personality and values of Ball:

Customer-comes-first attitude. "We are nothing without our customers," Colin Gillis, plant manager of Ball's beverage can joint venture—Rocky Mountain Metal Container—with Molson Coors in Golden, Colorado, points out. "We can be the greatest can maker in the world, but if we don't have someone to buy our products, we have failed."

Marty Ruffalo, senior vice president, beverage

Beverage cans whiz by on a high-speed production line. Ball makes beverage cans in North America, South America, Europe and Asia.

Every employee at the Williamsburg, Virginia, beverage can plant had a hand in earning the company's President's Safety Award. Safety is a top priority at all Ball locations.

container sales, adds, "Everybody in the company wants our customers to succeed."

Ball works extremely closely with its customers. At the Ball plant in Findlay, Ohio, for instance, every employee spends at least one day a year at a customer facility, seeing first-hand how the cans he or she makes at Ball are filled with soft drinks, beer, nutritional or other beverages. "It's really important, because it helps us understand our customers' problems and issues and builds long-term relationships," Jed Osborn, plant controller at Findlay, says.

The incredible power of people. "The greatest thing we have done is empower our people," according to Todd Farley, plant manager in Reidsville, North Carolina. "We have encouraged them to assume 'ownership' of their jobs and find ways to reduce costs and improve quality. That is a very powerful force."

To motivate employees and share with them the fruits of their efforts, Ball's incentive plans are not just limited to senior executives. "In the 1990s, we pushed our incentive plans right down to the shop floor in many locations," Seabrook says. As a result, these plans provide annual payments to employees based on meeting performance benchmarks.

Manufacturing skill and innovation. Ball is known throughout the packaging industry for its outstanding manufacturing capabilities, marked by low costs and high quality. "Operating superiority is a real key for us," John Friedery, senior vice president and chief operating officer, North American packaging operations, remarks.

Dave Hoover says, "We make our money in our plants. At the end of the day metal and plastic resin go into our plants and products come out, and that's the way we earn a profit. How our people do their jobs every day makes us successful."

Ethical conduct. Honest dealings with employees, suppliers and customers are a way of life at Ball, dating back to

Can-Do Attitude

Ball has many "owners"—not only shareholders but also employees who are encouraged to behave each day as if they owned the company. In fact, through stock ownership, the majority of Ball people do have a stake in the business.

Ball believes that employees who behave like owners provide the best value in products and services to customers and deliver superior results. To align employees' interests with those of shareholders, employees—like shareholders—share the benefits and risks of the company's financial performance. In a good financial year, Ball has paid as much as $60 million in incentive compensation to employees as part of their wages or salaries. In a bad year, it would not pay any incentive compensation.

Teamwork is a way of life at Ball. The company empowers its people to work together to solve problems and identify opportunities for cost savings and manufacturing process improvements. "Most of the equipment in our factories is not much different from that of our competitors," John Martin, senior vice president, purchasing, Ball Packaging Operations, says. "We outperform our competitors because of our people."

Steve Ferguson, engineering manager at the Fairfield, California, beverage container plant, notes that Ball's culture downplays egos and encourages cooperation. "Ball is a great place for someone who wants a supportive environment and a chance to assume responsibility quickly," he adds.

Teamwork, cooperation and collaboration are among the hallmarks of Ball's success. Left to right in photo at left, maintainers Thomas Nguyen, Robert Burns and Calvin Boone inspect a can at Ball's metal beverage container plant in Williamsburg, Virginia, in 1994.

Creating Shareholder Value

Ball directors, including CEO R. David Hoover, center, and director emeritus John W. Fisher (with gavel), celebrate the company's 125th anniversary by ringing the bell on the New York Stock Exchange on April 27, 2005.

If you had purchased $100 worth of Ball Corporation stock at the end of 1999, it would have been worth $472 at the end of 2004. The same $100 invested in the S&P 500 index would have been worth $88.

"One of the things we are most proud of is the shareholder value we have created," says R. David Hoover, Ball's chairman, president and chief executive officer. "The total value in the equity of our company increased from $1.1 billion in 1999 to nearly $5 billion in 2004. So for our 125-year-old company, 80 percent of its value as of early 2005 was created during the prior five years."

The company split its stock two-for-one in 2002 and again in 2004 and raised its dividend three times during the same period.

Few are more gratified by those achievements than Hoover, who is the face of Ball on Wall Street and remembers well a period when the company's stock price wouldn't budge.

"From 1987 to 1997 we made all kinds of changes and improvements within our businesses, and the stock price remained literally flat," Hoover recalls. "It was $8.84 split-adjusted at the end of 1987 and it was still $8.84 at the end of 1997. It was a transitional period that on one level was frustrating, but on another it made us stronger as a company and as a management team."

That experience clearly paid off for both Ball and its shareholders.

the values established by the company's founders. "This is not a place where you can succeed by cutting corners," says Joette Bailey-Keown, Ball's director of environmental services. Ball's Corporate Compliance Program, approved by the board of directors, emphasizes that "all employees are required to conduct the business affairs of the Corporation in a lawful and ethical manner." When faced with a business question that may be compliance related, employees are encouraged to ask themselves not only whether their decision and conduct will be lawful, but also whether it might create an appearance of impropriety and whether they could defend their action in public.

Straightforward Business Model

Ball employs the same business model in all its operations: earn more than it spends and generate cash while doing so. By rigorously following this model and investing in high-return business opportunities, the company has posted an outstanding growth record.

A primary objective of management is to align the interests of employees—and in particular senior managers—with those of shareholders and, in doing so, create stockholder value. The company has been highly successful in recent years in meeting this objective. Driven by the company's financial results, the stock price increased 554 percent from March 2000 through the end of 2004, one of the top performance records on the New York Stock Exchange for that period. As a result, even though Ball is 125 years old, more than 80 percent of its stock market value as of the end of 2004 had been created in less than five years, a remarkable development.

To be sure, there have been periods during the past 125 years when Ball has struggled. But the people of Ball have never wavered from the business practices and values—including customer service, teamwork, efficient manufacturing operations and integrity—that have made the company successful. Moreover, whenever the company has faltered financially, it has always found a way to revitalize itself and move on.

Ball is among a relatively small number of U.S. industrial corporations that have survived, let alone prospered, for as long as a century and a quarter. This has taken more than luck. Ball has endured and succeeded through the ingenuity and commitment of generations of Ball people, beginning with the five Ball brothers who founded the company in 1880.

Ball makes more than 50 billion two-piece metal beverage containers and tens of billions of easy-opening can ends for its customers every year around the world.

Exhibit
of the
1931 INTERNATIONL
CANNING CONTEST

CHAPTER TWO: 1880-1930

The First Half-Century: America's Home Canning Jar Company

Ball is best known by the public for a product it no longer makes—the Ball home canning jars used by generations of homemakers to preserve fruit, vegetables and other foods. Consumer surveys showed that the Ball name on a canning jar was as well known as the Ford name on a car.

Canning jars were one of the great utilitarian inventions of the 19th century. In 1810, a Frenchman named Nicolas Appert hit upon the idea of preserving food by heating it and sealing it in an air-tight container. However, it was left to an American named John Mason to develop the first glass jar with a screw top and rubber seal, making home canning practical. Mason patented his invention in 1858, and Mason "fruit jars," as they were called, became a national rage.

Ball got into the canning jar business in a roundabout way. The company was founded in Buffalo, New York, in 1880 by two brothers—Frank C. Ball, 23, and Edmund B. Ball, 25. They had grown up on an Ohio farm and were self-reliant and energetic. Prior to 1880, they had failed twice in business—first in the manufacture of "fish kits" (shipping boxes for salted fish), then with an oddly named rug-cleaning firm called The Elastic Steam Carpet Beating Works.

Determined to succeed despite those setbacks, in 1880 they borrowed $200 from their uncle George, a Baptist minister, and tried their hand at another business by purchasing the tools and materials of a nearly bankrupt tin can manufacturer. They named their new firm The Wooden Jacket Can Company.

Swinging quickly into action, the

Ball manufactured its first home canning jars in 1884 and sold them until 1993. Consistent quality helped make Ball canning jars the industry's best-selling brand and helped the Ball trademark become one of the most recognized in the nation.

Ball was founded in Buffalo, New York, in 1880. It initially made wood-sheathed tin cans for storing oil, paint and varnish. Two years later, the brothers who founded the company shifted into glass as the material for kerosene containers. Then, in 1884, they made their first glass fruit jars.

brothers rented a loft over a paint store, hung out a trade sign and hired a tinsmith to make wood-sheathed tin cans for storing oil, paint and varnish. Their first customer was a local paint and varnish dealer.

A few months later, at the urging of their widowed mother, Frank and Edmund brought their three other brothers—Lucius L., William C. and George A. Ball—into the firm as partners. The five Ball brothers were an unbeatable combination. They saw opportunities where others did not, and they had the business smarts to capitalize on them and grow their company.

In 1882, they hired a crew of glassblowers to make glass-lined containers for kerosene. Two years later, discovering that the patent for the screw-top fruit jar invented by John Mason had expired, they began producing home canning jars. That discovery would change their company and their lives.

Canning Jars: Ubiquitous in American Homes

Although John Mason received his patent in 1858, the manufacture of fruit jars didn't catch hold until after the Civil War. Consolidated Fruit Jar Company and Hero Fruit Jar Company were the first two major manufacturers. Bitter rivals, they dominated the market in its early years.

After Mason's patent expired in 1875, many new competitors entered the field. Newspapers of the 1880s and 1890s were filled not only with ads for fruit jars, but also with home canning recipes. There were many competing brands, and each producer and retailer claimed its jars were unsurpassed.

Ignoring the competing claims, the Ball brothers—late entrants to the business—cruised right by all the other manufacturers and soon became the premier supplier. In one of the extraordinary business success stories of the period, Ball emerged as the number-one fruit jar brand.

Muncie: Fruit Jar Capital of the World

In 1884, the year the Ball brothers made their first fruit jars, they renamed their firm the Ball Brothers Glass Manufacturing Company. Four years later, they moved the company from Buffalo to Muncie, Indiana. Because of Ball, Muncie became the fruit jar capital of the world.

The decision to relocate to Muncie was driven by access to plentiful supplies of cheap energy. Many factories of the era were fueled by coal, but the realization in the 1880s that natural gas—a less expensive fuel—could be tapped from wells and transported by pipeline set off an exploration scramble throughout the Midwest.

In 1886, Muncie became one of the first communities in Indiana to make a significant gas find. That same year, Frank Ball visited a glass plant in Findlay, Ohio, that used natural gas in its furnaces. He came away "deeply impressed" with the cost savings. He decided the time had come to move the company to a city with abundant gas supplies. One prospective location was Bowling Green, Ohio. While visiting there, Frank received a telegram from a man he did not know—James Boyce, a Muncie business owner and real estate developer—urging him to consider Muncie before choosing a site.

"I had never heard of Muncie," Frank Ball later wrote. "I looked it up on the map and found that it was a town of about six thousand population about fifty miles northeast of Indianapolis.... I decided to run down to Muncie and see what they had to offer." He liked what he found.

Real estate developers and business groups across the Midwest were bidding aggressively for new industry. In relocating to Muncie, the Ball brothers

The skills and motivation of employees have been vital to Ball since its beginning. The company was originally called the Wooden Jacket Can Co., changing its name to Ball Brothers Glass Manufacturing Company in 1884.

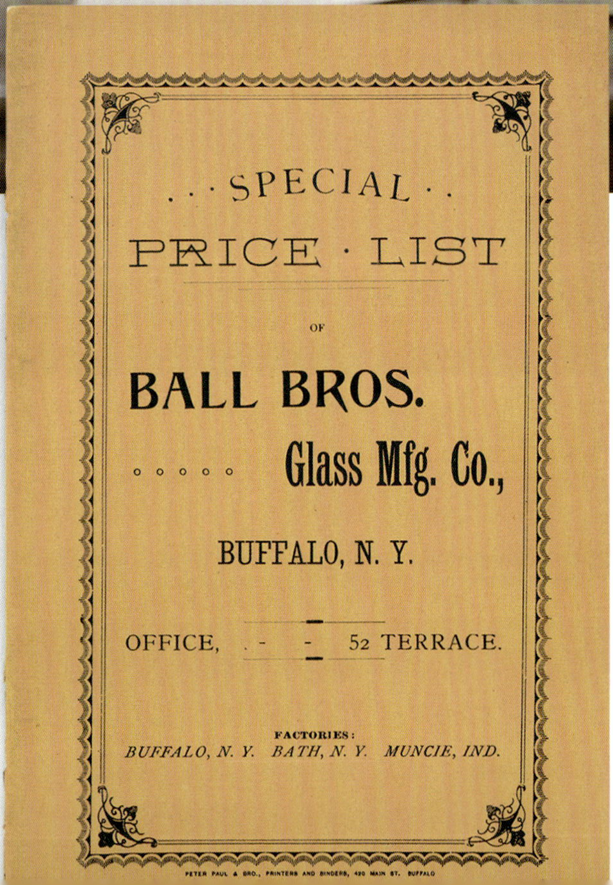

...SPECIAL..

PRICE · LIST

OF

BALL BROS.

Glass Mfg. Co.,

BUFFALO, N. Y.

OFFICE, 52 TERRACE.

FACTORIES:

BUFFALO, N. Y. BATH, N. Y. MUNCIE, IND.

PETER PAUL & BRO., PRINTERS AND BINDERS, 420 MAIN ST. BUFFALO

The five Ball brothers were, left to right, George A. Ball, William C. Ball, Frank C. Ball, Lucius L. Ball and Edmund B. Ball. They were not only capable businessmen, but also leading benefactors of the community. At right is an early price list for Ball products.

accepted an offer of $7,500 for moving expenses, seven acres of free land for a plant site, a railroad siding and free natural gas for five years.

Muncie was a small, remote community. When Ball's glassblowers arrived from Buffalo, one of them wondered whether he was still in the United States and whether he could mail a letter for the normal rate of two cents. All he knew, he said, was that he was "out West."

As it turned out, Ball and Muncie fit each other well. In less than a decade, Ball became the largest and most successful company in the city and one of the larger industrial firms in the state. Ball would remain in Muncie for 111 years.

Although John Mason had invented the screw-top home canning jar, the Ball brothers learned how to make jars efficiently and profitably. The brothers manufactured 17,500 jars in 1887, the year before they began production in Muncie. In 1892, they made nine million. By 1905, they were making 74 million jars a year and were supplying more than 90 percent of the home canning jar business in the United States, according to an article that year in the *Fort Wayne (Indiana) Sentinel.* Few companies in any industry grew so rapidly.

The Ball brothers became known in the press as the "fruit jar barons" or the "fruit jar tycoons."

Factors for Success

The brothers saw to every detail of their business, from technology to marketing. They succeeded for reasons that ring as true today as they did more than a century ago.

Most important, they were innovators who continually found ways to improve their products and reduce costs. When the Ball brothers entered the business, most jar manufacturers bought their zinc caps and rubber gaskets from independent suppliers. Because each component was manufactured by a different firm, the components did not always match, resulting in leakage that spoiled the contents. In 1888, the *Atchison (Kansas) Daily Globe* warned, "Fruit preserving season brings much hard work to the housekeeper, yet the wise woman will not fail to put up a good supply, for there are times when a jar of nice fruit is a 'friend in need.' But fruit jars are not all perfect. Therefore examine each one carefully and see that rubber and top fit so as to seal the jar hermetically."

The Ball brothers avoided this problem. Early on, they made the critical decision to manufacture their own caps and gaskets to ensure proper fitting. "Jobbers as well as housewives became aware of the fact that Ball jars were superior because of the seal," according to George E. Myers, a retired Ball financial executive, in his unpublished 1965 manuscript *The Story of Ball Brothers.*

Besides making a better product, the brothers were at the forefront of a technology revolution that brought automation to glassmaking. In 1898, they developed one of the first machines to automate some steps of the glassmaking process. In 1905, they pioneered an improved device to feed molten glass from the furnace to a mold. If the best technology was invented by others, the brothers were quick to adopt it. In 1909, they purchased the Greenfield Fruit Jar & Bottle Company to obtain its license to use the Owens automatic bottle making machine, which permitted dramatic cost reductions and production increases.

By investing in labor-saving technology, the company was able to reduce costs, trim prices, sell more products—and thereby increase employment even though operations were being automated. By 1892, the company had 1,000 employees; by 1905, it had 2,000.

Besides making a better product, the brothers were at the forefront of a technology revolution that brought automation to glassmaking.

The company also grew through acquisitions, purchasing smaller fruit jar firms throughout the Midwest. Communities welcomed the Ball brothers because of the jobs they created. After acquiring the Schram canning jar plant in Hillsboro, Illinois, in 1925, Ball Brothers modernized the facility and increased its workforce from 150 to 250. "The payroll amounts to more than $3,000 a week and will stimulate business wonderfully in Hillsboro," the *Decatur (Illinois) Review* declared.

The brothers never missed a beat when it came to maximizing the value of their plants. In 1912, they bought a zinc mill that was in receivership and moved it

ON

lock, stock and barrel from Kansas to Muncie. Because the mill produced more zinc sheet than they needed for jar lids, they used the excess capacity to make zinc casings for flashlight and radio batteries. Similarly, they used the excess capacity of their rubber gasket operation to make gaskets for refrigerators.

The Balls were shrewd marketers. Their firm was one of the first in any industry to put its name on its products. Their *Ball Blue Book* home canning guide and recipe book, first published in 1909, became indispensable to homemakers and helped spread the Ball name. In addition, the brothers pioneered the use of canning demonstrations to promote their products. The company had a network of demonstrators who attended meetings of local women's clubs to explain the whys and wherefores of home canning.

Fulfilling the American Dream

What were the Ball brothers like?

All five were "alert" and "capable" and "never spared themselves in business or civic affairs," in the words of a 1937 sociological study, *Middletown in Transition.* "Middletown" was the authors' code name for Muncie and the "X family" was their code name for the Balls. "In their modesty and personal rectitude," the study said, "combined with their rise from comparative poverty to great wealth, they fit perfectly the American success dream."

Frank, the fourth oldest of the brothers, had natural business talent and was the company's president and dominant personality for 63 years from its founding in 1880 until his death in 1943 at age 85. As he acquired

Ball moved to Muncie, Indiana, in 1887. The photo above shows an employee in the Muncie glass plant circa 1905. Glass jars, opposite, awaiting shipment are stored in a field south of Muncie in 1903.

Ball displayed its products, right, at the 1933-1934 Chicago World's Fair. The first edition of the famous *Ball Blue Book* in 1909, below, was titled *The Correct Method for Preserving Fruit*.

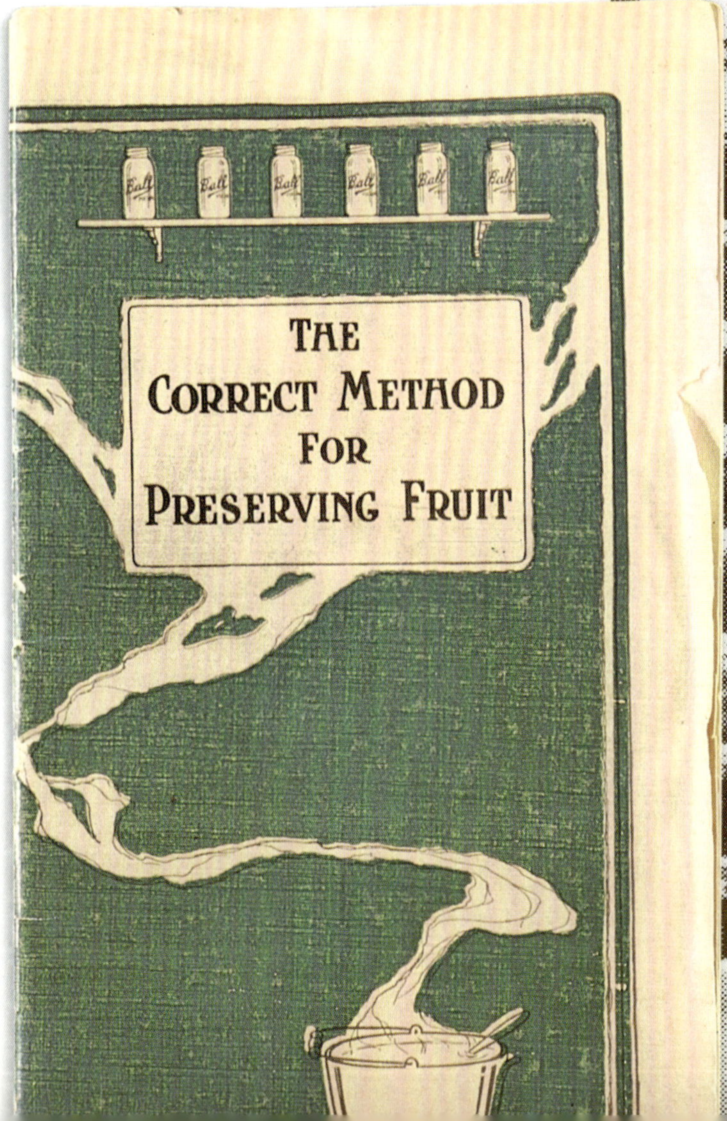

OTHERS CO.
Ball
JARS
BROTHE
Muncie, Indiana.
Glass Fruit Jars

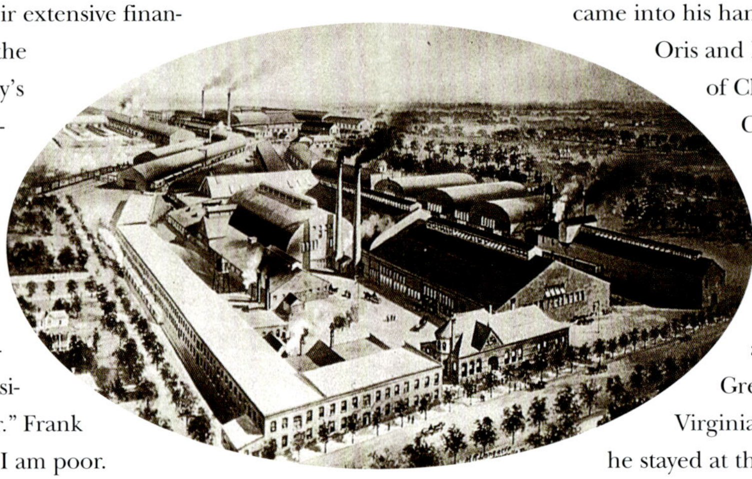

This early 20th-century birds-eye view of the Ball factory in Muncie is titled "The largest glass factory in America." Ball was attracted to Muncie by abundant supplies of natural gas for its furnaces.

wealth, he bought European art and eventually donated his collection to Ball State University (named for the brothers because of their extensive financial support), forming the nucleus of the university's galleries. He had a playful wit. In 1924, when Alvin M. Owsley approached Frank to ask for the hand of his daughter, Lucy, in marriage, Alvin supposedly said, "You are a business man; I am a lawyer." Frank nodded. "You are rich; I am poor. You're a Republican. I'm a Democrat." According to legend, Ball pulled at his moustache and responded, "I've known *some* honest lawyers, and I was poor longer than I've been rich. But now—about this Democrat business—I'm not quite so sure." Frank then welcomed Alvin into the family.

Edmund, the third oldest, was vice president and general manager, responsible for day-to-day operations. He loved machinery and often pitched in when problems arose on the factory floor. He was well-liked by employees, who presented him with a gold watch that became one of his prize possessions. When he died in 1925 at age 69, a Muncie newspaper commented, "He never asked a man to do something that he would not do himself."

William, the second oldest, was corporate secretary and worked mostly in sales, spending many hours on the road meeting with customers and prospects. He was an extrovert who entertained customers, employees and friends with a seemingly endless array of stories. He passed away in 1921 at age 68.

George, the youngest, was the company's treasurer. He collected rare manuscripts and books, and he had many private investments in the railroad, automobile and other industries. In the mid-1930s, he briefly became one of the nation's most powerful railroad magnates when the 23,000-mile Van Sweringen rail empire came into his hands upon the deaths of Oris and Mantis Van Sweringen of Cleveland, to whom George had provided financial backing. The empire consisted of 249 separate corporations, including seven major rail lines as well as the famous Greenbrier hotel in West Virginia. It was said that when he stayed at the Greenbrier and was assigned the presidential suite, he was surprised and impressed when he checked out and didn't have to pay—forgetting, perhaps, that he owned the place.

No Ball brother ever received more publicity than George did after he took possession of that agglomeration of railroads and other companies. His photo appeared in countless magazines and newspapers, and he was even called to testify before a U.S. Senate committee. A newspaper in Frederick, Maryland, carried his photograph with the caption: "Buys Rail Empire 'For Song:' George A. Ball tells a senate committee about it." Private and publicity shy by nature, he sold the railroads and other properties as quickly as he could. He died in 1955 at age 92.

Lucius, the oldest, left the company in 1887 to attend Buffalo Medical College, fulfilling his dream to become a doctor. He was quiet and thoughtful, with a gentle sense of humor. He continued on the company's board of directors even after he began his medical practice at age 40. He never owned a car and was a familiar sight in Muncie riding his bicycle to make house calls. He died in 1932 at age 82.

The five brothers lived next door to each other in beautiful homes on a bank overlooking the White River. They were extremely close and worked as a team, sharing responsibility for the management of their company.

Ownership of the firm was tightly held within the family. Incorporation papers from 1899 show that Frank and Edmund each owned slightly less than 37 percent of the stock, George 18 percent, William 6 percent and Lucius less than 1 percent. The remaining 2 percent was held by the brothers' older sister, Lucina, an educator.

Far-Reaching Charitable Activities

The brothers are remembered today not only for their business prowess but also for their charitable activities. They became Muncie's greatest benefactors, contributing generously to the YMCA, Ball Memorial Hospital, the American Legion and local churches, among other organizations. In 1922 alone, according to that year's *New York Times*, the brothers donated $1 million to charity, a remarkable sum for the time.

In 1917, the brothers purchased the campus of a defunct teachers' college in Muncie. The following year, they turned the property over to the state, which reopened the school. With ongoing financial support from the Ball family, the school evolved into today's Ball State University, which in 2004 had an enrollment of 20,500 undergraduate and graduate students. To this day, Ball family descendants are major benefactors.

In the 1930s, the Muncie Chamber of Commerce honored the brothers' generosity by commissioning a large public sculpture. The group selected Daniel Chester French—famous for his statue of Abraham Lincoln in Washington, D.C.'s Lincoln Memorial—to design the work. In a testament to the public's respect for the Ball brothers, more than 11,000 Muncie residents donated nickels, dimes, dollars and more to make the sculpture possible. The result is *Beneficence*, a large bronze sculpture of an allegorical woman, surrounded by five columns of Indiana limestone, one for each of the brothers. The statue stands on the Ball State campus, where it is known affectionately as "Benny."

Turning Point in History

The Ball brothers were among the most successful business owners of their era. They were not only accomplished businessmen; they also established many of the principles, including integrity, customer service and treating people fairly, that continue to guide Ball Corporation today.

By 1931, at the peak of the home canning jar business, Ball Brothers Company (its name had been shortened in 1922) was selling nearly 190 million fruit jars annually, equivalent to 1.6 jars per year for every man, woman and child in America. That same year, the company earned $5.5 million on sales of $14.6 million, both record highs.

Although the Great Depression was by now in its second full year, there was widespread hope that better times lay just over the horizon. Indeed, in August 1931, Ball reopened its fruit jar plant in Hillsboro, Illinois, which had been idle for nearly two years. "About 175 employees will return to work, which will be resumed at once," the *Associated Press* reported.

Unfortunately, the Depression got worse, not better. The Ball brothers had been riding a wave of endless growth for half a century, but that was about to change. Their company was entering one of the most difficult periods in its history.

The Ball brothers pose in an 1893 portrait typical of the formal style of the period. Their father was a farmer and inventor, and their mother a schoolteacher who urged them to go into business together.

CHAPTER THREE: 1930-1980

New Challenges, New Opportunities

Ball's first 50 years saw consistent, profitable growth in the fruit jar business and the development of important positions in the zinc and rubber products industries. Its next 50 years—beginning with the onset of the Great Depression of the 1930s—would be characterized by struggle, recovery and dramatic change.

Even during the worst years of the Depression, Ball remained the nation's premier fruit jar supplier. In 1938, the *Reno (Nevada) Evening Gazette* described the fruit jar as a "humble but important article of American domestic economy" and rhapsodized that the Ball brothers were "real servants of mankind who made home canning so easy that virtually every housewife, in spite of the vast array of ready-made delicacies on the market, does some preserving and canning at home."

But times were changing. Another newspaper, the *Helena (Montana) Daily Independent*, noted that demand for fruit jars no longer matched the peak levels attained "back in the days when home-canned fruits and vegetables lined the cellars of all frugal American housewives."

In fact, by the mid-1930s, demand was declining. Although Ball began the Depression on a positive note as sales of its canning jars reached an all-time high in 1931, that peak year was the calm before a ferocious storm. Sales plunged by nearly 50 percent to less than 100 million jars in 1932 and fell further in 1933 and 1934 as consumer demand collapsed. Many homemakers continued to can, but increasingly they were reusing screw-top grocery jars or recycling their old Ball jars. Although Ball would continue to manufacture canning jars for the next six decades, the

Ball's diversification included the acquisition of Rolled Plate Metal Company, a maker of zinc photo-engraving plates. Opposite, an operator at a plant in Greencastle, Indiana, inserts a zinc plate into a reversing, reducing mill—which thins, elongates and broadens the plate as it passes through. Ball's products for much of the century included glass bottles, left.

In 1927, Ball celebrated four decades of glass-making in Muncie with an imaginative parade float that featured a globe flanked by two oversized representations of canning jars.

Depression marked a critical juncture: canning jars were no longer the engine of the company's growth.

Despite a severe falloff in jar sales, Ball Brothers Company performed relatively well during the Depression by trimming costs and developing new glassware products to take up some of the slack. Amazingly, the company earned a profit every year in the 1930s, although barely so in 1938 when it made just $5,000. However, its factories were aging and its longer-term prospects looked increasingly dim.

Consequently, Ball embarked on a lengthy, sometimes tortuous quest to find newer businesses and modernize its facilities. It is never easy to turn a company in a new direction, especially a company that has relied so heavily on one product throughout its existence. Yet Ball Brothers' leaders did just that, saving their company from possible extinction and eventually making it bigger, better and more profitable than ever.

Early Diversification

There were no radical moves at first. The company diversified gingerly into related businesses, beginning with the manufacture of glass food jars for Kraft, Gerber and other consumer product firms. Following the repeal of Prohibition in 1933, the company also began making beer and liquor bottles, although Frank Ball, a teetotaler, referred to these containers as "special purpose" bottles or "flasks," assiduously avoiding words like "beer" and "whiskey." Whatever one wished to call them, by 1937 the company was selling more than 50 million liquor bottles annually.

Meanwhile, as their company coped with change, the Ball family remained an anchor of the community. As the Depression deepened, nearly 40 percent of the nation's banks failed between 1929 and 1933, but the Balls were not about to let that happen in Muncie. On February 27, 1933, at the height of the nation's banking crisis, Frank Ball calmed any banking jitters in Muncie by announcing that the family would supply $500,000 of collateral to the city's three banks to guarantee their stability. Because of the Balls, Muncie was an island of banking tranquility in a nation of financial turmoil. In 1933, the year prior to the inception of federal deposit insurance, depositors lost an estimated $540 million nationwide due to bank failures. Not a penny was lost in Muncie.

The Ball family also stepped forward when Muncie's largest department store, W. A. McNaughton, went bankrupt in 1933. To protect jobs and make sure Muncie continued to have a top-flight retailing presence, the two surviving Ball brothers—Frank and George—purchased the building, refurbished the interior and restocked the shelves. They reopened the facility under the name Ball Stores, which continued as a downtown Muncie institution for more than 50 years.

Noting their purchase of the store and widespread philanthropic activities, the *Ironwood (Michigan) Daily Globe* wrote, "In Muncie, Ind., with his elder brother Frank, George Ball is 'First Citizen.'"

World War II and a Major Antitrust Suit

Ball Brothers Company shifted to military production during World War II, making a variety of materiel including ammunition shells and producing zinc casings for more than half the batteries used by the American and allied forces.

In addition, the company promoted home canning as a way for housewives to support the war effort by conserving food and metal, the latter because home canning reduced the volume of store-bought goods in tin and steel cans. Newspapers across the nation rallied to the home canning cause. "Proud as a Peacock," the *Mountain Democrat and Placerville (California) Republican* wrote in 1944, picturing a homemaker holding a fruit jar filled with vegetables. "There's nothing to top the genuine feeling of pride that rolls up around kitchen shelves lined with jars of home-grown and home-canned foods."

That same year, Ball ran an ad in the *Ogden (Iowa) Reporter* picturing a soldier at the front lines and a

Ball promoted its home canning products with demonstrations for homemakers and advertisements in magazines and newspapers. The annual ***Ball Blue Book***, above, contained recipes and canning instructions.

LYNCH

Ball was an early leader in the automation of glassmaking. Automation, like that shown opposite, allowed the company to lower costs, reduce prices, sell more canning jars and increase employment. Ball-Bingham semiautomatic glassmaking machines, above, invented in 1900, were shipped by rail. The company made its own corrugated boxes, left, for shipping its products.

Frank C. Ball led the company for its first 63 years until his death in 1943 at age 85. He was a demanding, hands-on manager with a take-charge personality. In his autobiography, published in 1937, he said the company prospered "by practicing economy, perseverance, close application and faithful co-operation."

homemaker in her kitchen with the headline, "My heart is over there. That's why my hands are busy here." The ad encouraged Americans to "answer Uncle Sam's call" by canning more fruits and vegetables. Home canning surged during the war before resuming its decline after hostilities ended.

Throughout the war years, Ball coped not only with changing markets but also with a landmark civil antitrust case that was making its way through the federal courts. In 1939, the government sued Ball and eight other companies in the glassware industry. According to the government's complaint, Hartford-Empire Company—which made machinery for manufacturing glass—controlled the industry by restricting the licensing of its equipment. Frank Ball testified that, even though Ball Brothers didn't use Hartford-Empire equipment, the company had paid royalties to Hartford-Empire since 1933. The payments were made, he said, to stave off Hartford-Empire's threat of patent litigation. He also said he "understood" that Hartford-Empire would restrict the fruit jar production of two competitors if Ball paid the royalties.

In 1942, a federal judge ruled that all nine companies had violated federal antitrust laws. Because the case involved Ball Brothers, a name known in virtually every household in America, it made front-page headlines across the nation. The trial court's decision was upheld three years later by the United States Supreme Court. The courts not only enjoined Ball Brothers from acquiring any of its competitors or their plants (although it could still build new plants), they also ordered the company to sell its fruit jar factory in Three Rivers, Texas. Edmund F. Ball, Edmund B.'s son, later said the decision, as difficult and embarrassing as it was, may have been "a blessing in disguise" because it forced the company to diversify into new areas with growth potential.

The Next Generation

In 1939, when the antitrust suit was filed, Ball held 55 percent of the nation's fruit jar market. Kerr Glass Manufacturing Company, founded in 1903 in Portland, Oregon, held 22 percent. Other manufacturers, including Hazel-Atlas and Owens, supplied the remaining 23 percent of the market.

Despite its large market position, Ball was beginning to struggle financially as its outdated facilities and aging management took their toll. However, the public and most employees knew nothing about the company's financial problems since results were kept strictly private. Ball continued to be widely admired for its quality products and customer service, and outsiders had no idea that profits were lagging.

Frank C. Ball had led the company since its founding

Ed Ball, Gentle Visionary

Warm, unpretentious and even a bit shy, Edmund F. Ball led Ball Corporation for nearly two decades and revitalized its operations after World War II. In 1995, a company publication said, "Mr. Ball is Ball Corporation's Walt Disney, a man employees seem to universally respect and genuinely like and a symbol of the kind of company his father and four uncles founded in 1880."

Ball was a son of cofounder Edmund B. Ball. Born in 1905, he joined the company in 1928 after graduating from Yale, leaving in 1941 for military service. He saw combat in Italy during World War II and advanced to the rank of major before returning to the family firm in 1945. Serving as CEO from 1948 to 1963 and 1967 to 1970, he rejuvenated the business by modernizing its glass factories and diversifying into new fields such as aerospace and beverage cans. Equally important, he embodied the values that have characterized the company from its founding, including integrity, social responsibility and hard work. "Reputation is a tremendous asset," he once said. "I don't ever recall having to apologize for being associated with Ball Corporation or its employees, and I don't expect I will."

Edmund F. Ball turned the then-struggling company around after becoming Ball's president in 1948.

He maintained a deep personal interest in the company even after he retired, and he loved to visit its facilities and chat with employees. Employees in Findlay, Ohio, still remember when he dropped by in 1995 with his wartime friend, Lord David Hunt, on a Sunday afternoon. Hunt was about to return to England, and Ball—full of pride about the company—wanted to show him a Ball Corporation factory. On arriving at Findlay, Ball took Hunt onto the factory floor to chat with employees. "You could see how thoroughly Ed Ball enjoyed himself," plant manager Tom Martin says. Before they left, Ball and Hunt and their wives were honored at a hastily arranged reception in the plant conference room at which several employees spoke about the impact of Ball Corporation on their lives. Clif Reichard, a customer relations consultant and former vice president of glass container sales, says, "I had been with Mr. Ball on many occasions, both sad and happy, but this was the first time I had ever seen him overcome with such emotion to be moved to tears."

Ball never lost his zest for life, remaining physically and mentally active well into his 90s. He earned his pilot's license in 1930 and continued to fly until his final years, always taking along a Ball fruit jar as a good-luck charm. He began scuba diving at age 67, approaching it with the same vigor with which he approached every other pursuit. He even wrote two articles about the joys of scuba diving for the *Saturday Evening Post*. At age 91, he and his wife, Virginia, went to the North Pole aboard a Russian icebreaker. Arriving at the "top of the world," he donned a wet suit, hauled out a collapsible bicycle and did his version of a triathlon, riding, running and swimming around the pole.

Ball shared his financial good fortune with others, donating to an array of nonprofit causes, including Ball State University. He had a particular interest in telecommunications and public broadcasting. Fittingly, the Edmund F. Ball Building at Ball State houses the public broadcasting operations and telecommunications department.

When Ed Ball died in 2000 at age 95, Ball State President Blaine Brownell said, "He was a remarkable man who lived a remarkable life. He moved three times faster than the rest of us."

John W. Fisher, third from the left, succeeded Edmund F. Ball as CEO in 1970 and led the company for 11 years. During his tenure, sales increased from $170 million in 1970 to $815 million in 1981. On a trip circa 1958 to Barton Distilling Company in Bardstown, Kentucky, Fisher—then vice president of manufacturing—was joined by several Ball and Barton executives. The meeting included Barton President Paul Krouse, at left; Charles Warren, Ball glass division sales and service manager, second from left; and Jack Bellinger, vice president of Ball's glass division, at right.

six decades earlier. He was astute and capable and was as responsible as any individual for its many years of success. "F. C. was a very dominating guy" who oversaw every decision, according to John W. Fisher, who married Frank's niece and headed Ball Corporation from 1970 to 1981. Even after he suffered a stroke in 1939, Frank "continued to be involved in every stage of the company's operations to the extent his health permitted," according to Earl L. Conn in his 2003 book *Beneficence: Stories About the Ball Families of Muncie.*

However, some thought that Frank had become set in his ways and was reluctant to invest in new facilities or diversify outside the glass business. When Frank died in 1943 at age 85, his brother George succeeded him at age 80. Financial results worsened as the war ended.

George was a caretaker who led the company until 1948, when the next generation of Balls finally got its chance. Edmund F. "Ed" Ball, the son of cofounder Edmund B., had joined the company in 1928 and had been vice president in charge of glass operations when he left for World War II army service. He achieved the rank of major before returning to civilian life in 1945. Ed had a lively mind and warm personality, and always took great pride in the family business and the values it stood for. Tom Martin, manager of the Ball plant in Findlay, Ohio, describes Ed Ball—who died in 2000 at age 95—as "very down-to-earth and just a wonderful gentleman."

For a while during the postwar period, the family thought about selling the company. But family members asked Ed Ball if he wanted to become president and try to restore its profitability, and he said yes. In January 1948, at age 43, he was elected president and chief executive officer.

Ed Ball was an avid pilot who was fascinated by technical advances in aviation. He returned to Muncie following World War II convinced of two things, according to Robert Arentz, a longtime Ball Aerospace employee: "first, that his glassmaking business was dying for lack of modern technologies and manufacturing techniques; second, that he had seen the power of modern electronics-based technology first-hand in aircraft, and therefore knew the power latent in technologies in general." One of his early decisions was to hire Arthur Gaiser as director of research and development. Gaiser had earned his bachelor's and master's degrees from Alfred University and held more than 40 U.S. patents. Among other innovations, he developed glass capable of conducting electricity. Although his background was in ceramics engineering, he was full of ideas for improving glassware manufacturing and strongly favored investing in new technologically oriented businesses such as aerospace.

Before those investments could be made, however, Ed Ball faced the immediate task of reviving a celebrated company with a pulse that was barely detectable. In 1948, the year he became president, American industry was booming. The earnings of the blue-chip Dow Jones Industrial stocks surpassed the 1929 level for the first time, but Ball lagged, earning only $330,000 as if it were still stuck in the Depression. The following year, most corporations improved their earnings further. Ball, by contrast, lost money for the first time in its 64-year history, although only the board of directors, senior management and the company's auditors were aware of this dire fact.

At first, there was great resistance to change within the company. Employees were proud of Ball and its heritage and believed it was still generating strong earnings. Many could not understand why they suddenly needed to pay attention to costs and profits. They wondered why the Ball family, noted for its beneficence, didn't invest some of its wealth in the company to keep the business going regardless of its profitability. Ed Ball recalled, "I was shocked when discussing the company's problems with a key employee whom we considered might be qualified for more important positions in the organization. I was emphasizing the importance of profits and he responded, 'I've never had anything to do with profits and never will!'"

Years later, Ed Ball wrote, "Ball Brothers Company was struggling for its existence and desperately needed something, almost anything, to improve its likelihood for survival."

Ed Ball had seen the use of technology in World War II and was an avid pilot who was fascinated by technical advances in aviation.

Revitalizing the Company

Family companies often succumb to bad decisions, lack of fresh management or family disputes. Ball did not. At its most desperate hour, the people of Ball rallied around their company and took the difficult actions required to ensure its future. The company's revival began in the 1950s and took hold in the 1960s, when revenues and earnings rebounded. Ball Brothers Company, once all but dead, staged an amazing recovery.

For starters, Ed Ball and his management team modernized the company's glassmaking capacity by building new plants in El Monte, California; Asheville, North Carolina; and Mundelein, Illinois. With these new factories up and running, Ball closed its antiquated glassmaking plant in Muncie, a difficult decision that caused pain and criticism in the community. Ball's last glass container in Muncie—a city that had grown with Ball and thrived on its presence—was manufactured on March 19, 1962. "It was an emotional moment for all those present," *Ball Line*, the company's employee publication, reported.

"Although there was no outward display, the silence and sober expressions betrayed the feelings of all."

The company's rubber products business also was doing poorly, so Ed and his team acquired Chardon Rubber Company in Ohio and placed all rubber operations under Chardon's management.

At the same time, the company began to diversify. In 1951, it entered a new business by acquiring Kent Plastics Company, which operated an injection molding and decorating factory in Evansville, Indiana.

In 1956, Ball became an early corporate player in high technology when it bought Control Cells, a tiny Colorado firm that made an experimental device that Ball hoped to use to weigh ingredients more precisely in glassmaking. Ball's board of directors rejected the purchase, so Ed Ball—who was determined to develop a high-tech presence for Ball—wrote a personal check for $10,000 to buy the company. John Fisher, who was a member of the board and favored the purchase, says, "During the course of the next month, we talked to the other directors and at the following board meeting the purchase was approved and Ed's money was returned to him."

The weighing device was a commercial failure, but Control Cells—renamed Ball Brothers Research Corporation and today called Ball Aerospace & Technologies Corp.—soon acquired rights to a "pointing control" developed by scientists at the University of Colorado (CU). This control, when installed on a rocket, spun in the opposite direction from the rocket's engine, providing stability that enabled the rocket to be pointed accurately at a fixed point in space. "The Army and Navy and the Air Force just loved the CU pointing control," according to Bob Arentz. He believes that every spacecraft Ball Aerospace has ever produced "can trace its design heritage in one way or another" to that highly innovative device.

The company's aerospace business grew rapidly, fuelled by the American-Soviet space race. By the 1960s, Ball Brothers Research was participating in important contracts for the National Aeronautics and Space Administration (NASA), including the building of NASA's second science satellite, the Orbiting Solar Observatory (OSO), in 1962. Nonetheless, Ball's aerospace work provoked widespread skepticism and some unusual comments. In 1962, referring to OSO, *Business Week* wrote:

> *From Cape Canaveral last March, the National Aeronautics and Space Administration launched a sophisticated scientific satellite built by Ball Brothers Incorporated, and the newspapers made sport of the space work of the "fruit jar firm".... It wasn't the kind of company that's generally associated with Cape Canaveral or the highly technical, multi-billion-dollar U.S. space effort. Yet its subsidiary, Ball Brothers Research Corporation of Boulder, Colorado, little more than five-years-old, has managed to climb to 20th place in last year's list of NASA contractors.*

In 1965, Ed Ball coined the phrase "from fruit jars to satellites" because "I was still trying to convince people we just weren't making fruit jars, we were doing a lot more."

Diversification continued. In 1960, the firm acquired Rolled Plate Metal Company, which made zinc photoengraving plates for printing newspapers and magazines. The following year, the *Herald-Press* of St. Joseph, Michigan, reported that Ball "now has plants in 18 states manufacturing glass, zinc, graphic arts materials, mechanical products and rubber goods." Subsequently, the company diversified into computer disc drives, prefabricated housing, acrylic giftware and other areas.

However, no acquisition was more important than the 1969 purchase of Jeffco Manufacturing Co., which made metal beverage cans in Colorado. John Fisher recalls that Jeffco was "in the midst of developing several ideas [for improving the production of cans] and we encouraged them to go ahead with those ideas." In turn, Ball had significant expertise in metallurgy because of its zinc operations and was able to use that expertise to help Jeffco speed its production lines, reducing costs. The purchase of Jeffco was the first in a series of investments that eventually catapulted Ball to the top of the beverage can business in North America.

Ball also poured money into its existing businesses where it saw opportunities. In 1970, it completed a modern zinc plant in Greeneville, Tennessee, at a cost of $12 million, the company's largest single investment to that time. Capitalizing on the plant's low-cost operations and technology leadership, Ball became the number-one supplier of battery cases, coin blanks and other zinc products in the United States. "We were able to make zinc products at a price none of our competitors could touch," Fisher says.

In less than 20 years, Ball had modernized its facilities and made a remarkable transition to newer fields. In 1969, the company dropped the word "Brothers" from its name, becoming Ball Corporation. Ed Ball explained the move by saying that Ball Brothers was appropriate "when we were a small Indiana firm making fruit jars for home canning... but the name Ball Corporation more accurately conveys the scope of our expanding interests." Because of diversification, fruit jars—once the company's dominant business—now accounted for less than 10 percent of revenues.

Ball Aerospace has participated in important NASA contracts since the early 1960s. In 1973, four Ball Aerospace-built instruments were used on Skylab, left, to study the sun. The space shuttle Endeavor, opposite, lifts off on December 2, 1993, carrying the company's Corrective Optics Space Telescope Axial Replacement (COSTAR) unit to fix the Hubble Space Telescope.

Ball was family owned until July 1972, when it sold shares to the public for the first time. It listed its shares on the New York Stock Exchange in December 1973. President and CEO John Fisher, left, and Vice President of Operations Richard Ringoen pose on the exchange floor during a visit to New York. Ringoen succeeded Fisher as CEO in 1981.

Yet, even as it grew and diversified, the company retained its feel as a family business. Dave Hoover, today's chairman, president and CEO, recalls being surprised by two things when he visited Ball in 1970 for a job interview: He was greeted at the door by the company's most senior executives, and the offices were unusually modest.

Becoming a Public Company

In 1970, John Fisher succeeded his brother-in-law, Ed Ball, as CEO. Sales that year were $170 million. Fisher was a capable, tough-minded executive who joined the company in the 1940s after receiving a bachelor's degree from the University of Tennessee and an MBA from Harvard Business School. On taking charge in 1970, he continued the company's expansion while emphasizing strategic planning as an essential tool for any well-managed corporation.

At that time, the nation was beset by "stagflation": an unusual combination of slow economic growth, rising interest rates and accelerating inflation. The inflation rate jumped from 5.9 percent in 1970 to 10.4 percent in 1975 and 13.6 percent in 1980, creating headaches for consumers and businesses alike. As the nation's economy faltered, corporate bankruptcies increased sharply in the 1970s.

Under Fisher's leadership, Ball overcame these conditions and increased its revenues and earnings to new highs. He began by strengthening the company's balance sheet to provide capital for investment in profitable opportunities. In 1971, Ball refinanced its debt, and the following year took the historic step of selling 22 percent of the company's stock to the public. Until then, the company had been owned exclusively by members of the Ball family. "Ball Corp.'s initial public offering of 1,002,284 common shares sold out after reaching the market at $26 each," the *Wall Street Journal* reported on July 14, 1972. One year later, the company listed its shares on the New York Stock Exchange.

Fisher explains that Ball, as a public company, was better able to raise capital and acquire other businesses. In the early 1980s, he noted, "We'd been frustrated in some of the acquisition possibilities. We passed up a number of opportunities because we didn't have a publicly owned common stock to offer."

Following its public offering, Ball continued its growth by opening new facilities and making acquisitions. Having acquired Jeffco and its beverage can plant in 1969, the company grew its container business by building plants in Williamsburg, Virginia; Findlay, Ohio; and Fairfield, California, in the 1970s. It expanded in rubber and plastics by building a facility in Fort Smith, Arkansas; buying one in Batavia, New York; and acquiring Unimark Plastics of Milroy, Pennsylvania.

In addition, it expanded its commercial glass operations by purchasing Metropak Containers from Kraft Inc. in 1980 for $38 million. The Metropak deal fit Ball's strategy of increasing its sales of glass jars and bottles to food companies such as Kraft. The purchase doubled the number of Ball glass plants to eight and broadened its geographic spread, making the company more competitive in supplying companies with regional processing activities.

To expand the capabilities of Ball Aerospace & Technologies Corp., Ball acquired instrumentation maker Time-Zero Laboratories of Gardena, California, and the environmental monitoring business of Sierra Research Corporation. Ball also diversified into entirely new fields, such as agricultural irrigation systems. To provide focus, Fisher developed the idea that Ball is primarily a packaging company with a technology base, a concept that remains true to this day.

In 1980, Fisher said, "It may be self-serving to say we know what we are doing, but we really do. We plan. We seek out sound, growing markets. We organize. We anticipate. We budget carefully. We work hard. We are productive. We manage change, rather than let change manage us. If that is special or unusual, then perhaps we have special or unusual management skills." That same year, he was elected chairman of the National Association of Manufacturers, reflecting the esteem in which Fisher and Ball Corporation were held.

"Plenty on the Ball," *Financial World* magazine proclaimed when it profiled the company in 1980. Ball was not only enjoying its best results ever, but also would continue its growth well into the 1980s before facing—and overcoming—another of the difficult periods in its history.

In 1975, in an early attempt at international expansion, Ball began marketing turnkey irrigation projects for agricultural development in the Middle East. The division was phased out in the mid-1980s.

Ball

CHAPTER FOUR: 1980-2005

Good Times, Tough Times, Great Times

John Fisher retired as CEO in 1981, having led the company through 11 years of consistently strong growth. Earnings hit a new high in 1981 for the fifth consecutive year, while sales of $815 million were more than quadruple the level when he took charge in 1970.

Fisher had turned Ball into a large, publicly owned company with modern production facilities. *The Insiders' Chronicle*, an investment newsletter, noted in June 1981 that Ball was a leader in applying technology and automation to the metal and glass container businesses. It quoted a Wall Street analyst as saying, "They are surely one of the most efficient, low-cost packaging suppliers in the industry."

Fisher continued as chairman and a member of the board of directors until 1986, when he became chairman emeritus.

Continuing Ball's Growth

Fisher was succeeded as CEO by Richard M. "Dick" Ringoen, 55, whose background was in engineering and science. Ringoen was born in Iowa and thought about becoming a minister, banker or lawyer before enrolling at the University of Iowa, where he studied electrical engineering and earned his bachelor's and master's degrees. He started his career as a research engineer at Collins Radio Company and then spent 11 years at Martin Marietta Company, where he was involved in the development of the Titan intercontinental ballistic missile.

In 1970 (the same year Fisher became Ball's CEO and that current CEO Dave Hoover joined the company), Ringoen was hired by Ball Aerospace as a vice president and project manager. Even though Fisher was new in his position, he began looking almost immediately for a successor "because I felt that the most important assignment of a chief executive officer is to be sure he has someone to back him up." He was seeking an individual not only with leadership skills, but also with a technical background in the belief that Ball's future lay

Our Centennial Year Ball 1980

A cover from Ball's marketing literature around 2001, opposite, showcased its global reach and range of products, from beverage cans and plastic bottles to spacecraft. The company created the logo at left to celebrate its 100th anniversary in 1980.

Dick Ringoen was named CEO of Ball in 1981, succeeding John Fisher. The October 1981 issue of *Indiana Business* heralded Ringoen, who had joined the company 11 years earlier as a vice president of Ball Aerospace, as "A Man Who Knows How to be Number One." He retired in 1991, having led the company through one of its most dynamic periods ever, featuring significant expansion in the beverage can market, the development of international business and growth in its plastics and aerospace operations.

in the application of technology in its operations. Fisher looked inside and outside the company for more than two years before getting to know Ringoen and identifying him as a prime candidate.

In 1974, Ringoen was named Ball's vice president, operations, moving from Boulder to Muncie. In 1978, he was promoted to president and chief operating officer and three years later succeeded Fisher, leading the company for the next 10 years.

Ringoen, who died in 1993, brought his own special management style to Ball. He was personable and well-liked and had great fun running the business. Employees responded to his openness, enthusiasm and deep commitment to the Ball principles of integrity and teamwork.

Retired President and Chief Operating Officer George Matsik tells about the time he was transferred from managing a plant in California to running one in Virginia. "I came in for a monthly performance review meeting," he recalls, "and I said, 'Dick, I need almost a million dollars for some new printers to improve this plant's results. Without new printers, we just can't run the production speeds we need to run.' He replied, 'You know, George, when I was a youngster, every Christmas I'd come downstairs and all I wanted was a Lionel train. And every year for four or five years I'd come down and look everywhere and there was no train.' I said, 'What's the point of that, Dick?' He said, 'The point, George, is there are some things in life you never get and you're not getting your new printers.'"

Matsik continues, "About six months later, just before another performance review, I went out and bought a fancy Lionel train set and wrapped it as gift. All the senior people were at the meeting, and here I was just a plant manager. But I walked in with this big gift and presented it to Dick. When he opened it, he roared with laughter for what seemed like 15 minutes. Then he turned to me and said, 'You've got your printers, buddy. Go buy them.'"

1980-1987: Continued Profitable Growth

Ringoen inherited a company that was performing like a well-oiled machine. He continued its growth, raising sales above $1 billion for the first time in 1984 and increasing earnings to $60.1 million in 1987, the 11th consecutive yearly profit increase. Fueled by this performance, the company's stock more than quadrupled in price from 1976 through mid-1987. The company split its stock twice—in 1982 and again in 1985.

Ringoen took many actions to reduce costs, increase revenues and improve earnings. Most important, the beverage can market was in its golden age of growth: demand in North America increased from 55 billion cans in 1980 to 77 billion in 1987 and then to 103 billion in 1994. Under Ringoen's leadership, the company built beverage can plants in Saratoga Springs, New York, in

Ball entered the PET plastic bottle business in 1994. Today it manufactures approximately six billion plastic bottles a year for soft drinks, bottled water and other consumer products in a wide variety of shapes and styles, as shown at left.

1981 and Conroe, Texas, in 1987 and acquired a plant in Tampa, Florida, in 1988. With the Tampa acquisition, the company had a total of seven beverage can plants across the United States.

He also continued the development of the company's international business, including the creation of a joint venture with MC Packaging (Hong Kong) Ltd. and others in 1986 to build a plant in Guangzhou, China. He pointed out at a meeting of investment analysts that the plant was China's first two-piece beverage can facility. "If it fails we've lost $2 million," he said, "but if it succeeds, my gosh, what a market."

Because of his technical background, Ringoen was highly attuned to opportunities in aerospace. In 1985, he announced that the company planned to double the revenues of its aerospace business by 1990 and said Ball would be "buying its ticket to the 21st century" by doing so.

He also sought to grow the company's plastics business. Unlike Ball's plastic bottle operation today, this earlier plastics business made food containers and injection-molded plastic parts. Although Ball correctly identified plastic containers as a growth opportunity, this early venture was plagued by technical problems and never fulfilled its promise.

Ringoen's management style and success in growing the company won recognition among investment analysts and in the financial press. In 1985, *The Wall Street*

For many years, the Ball name was synonymous with glass packaging. The company not only made home canning jars for the consumer market, but also manufactured bottles and jars for companies such as Campbell Soup, Kraft Foods, Ocean Spray and Sebastiani. Glass jars, opposite, were made at the Dunkirk, Indiana, factory, one of 13 commercial glass plants the company operated before Ball sold its glass operations in 1996.

Transcript named him "the best chief executive in the container industry." It cited his "carefully calculated, long term strategy" and said Ball had "put together a better record than any other company in the industry over the last ten years."

Three years later, *Forbes* honored Ball as the nation's "most innovative packaging company."

Then the tough times arrived.

1987-1997: Decade of Challenge

The period from 1987 through 1997 saw nearly everything go wrong for Ball that possibly could go wrong. Dave Hoover refers to these years as a "period of change and preparation" and notes that the company made many attempts to create value for shareholders—some successful, some not. None of them, however, bolstered the stock price or led to the kind of sustained earnings growth the company had enjoyed from 1977 through 1987. It was not until the 1998 acquisition of the Reynolds Metals Company can plants that Ball began to deliver consistently stronger earnings and better stock market performance.

The period began on a distinctly sour note when the Dow Jones Industrial Average plunged a record 508 points, or nearly 23 percent, on October 19, 1987. Ball stock had already fallen 16 percent since late September 1987. It dropped an additional 24 percent on October 19—and then just sat there, not recovering. For the next decade, "the stock was flat like a dead man," Hoover recalls.

Profits also began to suffer. After reaching a new high in 1987, per-share earnings weakened and did not return to the 1987 level until 1991, after which they fell again.

Long Journey to Divest the Glass Business

Earnings got an initial jolt when the price of aluminum skyrocketed, adding to the cost of making beverage cans. Between 1985 and 1988, the average price for primary aluminum more than doubled.

Ball's commercial glass business also began to underperform. Ringoen told *Forbes* in 1988 that glass packaging—which accounted for approximately one-quarter of Ball's revenues—was "the foundation of the company and we'd better not play too loose here" in deciding how to improve its returns. Ball was then the nation's fourth largest supplier of food, juice, wine and other consumer-product glass bottles. It began a nearly decade-long struggle to get bigger and more competitive in the face of industry overcapacity, weak prices and the trend away from glass packaging toward plastic and metal.

Like all packaging companies, Ball was straining to find the right answers. "The packaging industry was like a game of gin rummy in 1987 as players swapped, drew and discarded glass, metal and plastics businesses, each certain that its combination would make it the packaging leader of the future," *Forbes* wrote in January 1988.

In 1987, Ball merged its commercial glass operations with those of InCon Packaging, which was owned by Holland-American Investment Corporation. Ball and Holland-American became 50-50 joint-venture partners in the new firm, called Ball-InCon Glass Packaging Corporation. The venture was the nation's third largest producer of glass containers. Because of its size and economies of scale, Ball-InCon was able to reduce costs and improve earnings. Ball was pleased with this progress and bought out its partner for $65 million in 1990, becoming the venture's sole owner.

But overcapacity still plagued the industry. Recognizing the need to continue to get bigger and more efficient, Ball acquired the commercial glass operations of Kerr Group, Inc. for $68 million in 1992, merging them into Ball-InCon. Ball closed two of the four plants acquired in the Kerr deal and replaced the majority of the lost capacity with modern expansions at other Ball facilities. However, market conditions deteriorated further, and financial results remained unsatisfactory. Ball subsequently closed its glass container plants in Asheville, North Carolina, and Okmulgee, Oklahoma, and took other steps to reduce costs and take capacity

out of the market. Financial results still lagged.

In 1995, Ball decided to get out of the glass business. The company welcomed the opportunity when Compagnie de Saint-Gobain of France offered to form a joint venture with Ball. Ball accepted the offer, contributing its glass business to the venture and receiving $145 million in cash. Saint-Gobain put up capital. The joint venture then borrowed money and bought Foster Forbes Glass Co., at the time owned by Pechiney of France. The combined new company was called Ball-Foster Glass Container Co. Ball Corporation owned 42 percent—Saint-Gobain owned the rest—of what was the second largest commercial glass maker in the United States.

"The deal is remarkably deft," *Financial World* said. "Ball's glass business is worth much more to a motivated strategic buyer like Saint-Gobain than it would be to just about anybody else."

The deal included a series of options that allowed Ball to sell its 42 percent Ball-Foster stake to Saint-Gobain. Media reports suggested that Ball would exercise its option and complete its exit from the glass business by 2000. As it turned out, however, Ball was able to accelerate the timing when Saint-Gobain sought to acquire Anchor Glass Container Corp. in 1996 and merge it with Ball-Foster to create the nation's largest glass company. Ball, which did not want to get bigger in the glass business, faced a choice: either invest additional money in Ball-Foster or see its 42 percent ownership diluted by the Anchor acquisition.

"To invest more would not be prudent," Dave Hoover told the *Indianapolis Star and News* in 1996. As a result, Ball and Saint-Gobain were at loggerheads over whether Ball-Foster should acquire Anchor Glass. To resolve their differences, Saint-Gobain agreed to purchase Ball's 42 percent stake for $337 million. Explaining why Ball accepted the offer, Hoover said, "We saw an opportunity to get a good price. We've taken the uncertainty out. We have the cash." The $337 million price, combined with the earlier $145 million received from Saint-Gobain, gave Ball a total of $482 million from the sale of its commercial glass operations, more than many Wall Street analysts had anticipated.

Ball had ended its long association with glassmaking.

The Frustration of Change

Although the company was able to sell its commercial glass operations on attractive financial terms, other moves did not work out as well, at least in the short term. Even those that created value did nothing to improve the stock price. The company kept looking for opportunities, yet was often frustrated by the results.

In 1988, Ball expanded its beverage can operations and entered the steel food can business by forming a joint venture in Canada, but financial results were unsatisfactory. The venture lost $2.1 million in 1990 and went into receivership. The following year, Ball acquired full ownership of the venture out of receivership and was eventually able to make it profitable.

In 1990, Ball agreed to acquire the beverage can business of Continental Can Europe for $1 billion. The proposed deal would have energized Ball by increasing the size of its beverage can business and giving it a large position in the growing European market. However, Ball was unable to complete the deal because of financing issues.

In 1992, the company expanded in the People's Republic of China by investing in a newly formed joint venture, FTB Packaging Ltd., to build plants there. Five years later, when it purchased a controlling interest in M.C. Packaging (Hong Kong) Limited, Ball was involved in more than 20 metal and plastic packaging plants in China and was that nation's largest beverage can manufacturer. However, Chinese demand for beverage cans began to weaken along with the Asian financial markets, causing Ball to scale back operations and take write-offs in 1998, 1999 and 2001. Hoover says Ball was too enthusiastic about the growth prospects of the Chinese beverage can market and didn't fully understand the risks. Today, Ball has a smaller, profitable business in China producing beverage cans and ends as well as plastic containers for the oil industry.

In 1993, Ball acquired food can manufacturer Heekin Can, Inc. and its 12 U.S. plants, despite a consultant's advice not to do so. It struggled for several years to achieve adequate profitability.

In 1994, the company tried to sell Ball Aerospace, soliciting bids. However, it could not find a buyer willing to pay more than $150 million. Unwilling to accept such an offer, Ball kept the business. The following year it sold a non-core unit of Ball Aerospace, the Efratom time and frequency measurement business, to Datum Inc. for

Ball Aerospace designed the Combined Release and Radiation Effects Satellite (CRRES), launched by NASA in 1990 to study the electrical, magnetic and particle environment of space near Earth. CRRES flew in the harshest of orbits and measured the impact of cosmic bombardment on the electronic instrumentation of satellites. The data it gathered helped scientists design better satellites that function longer.

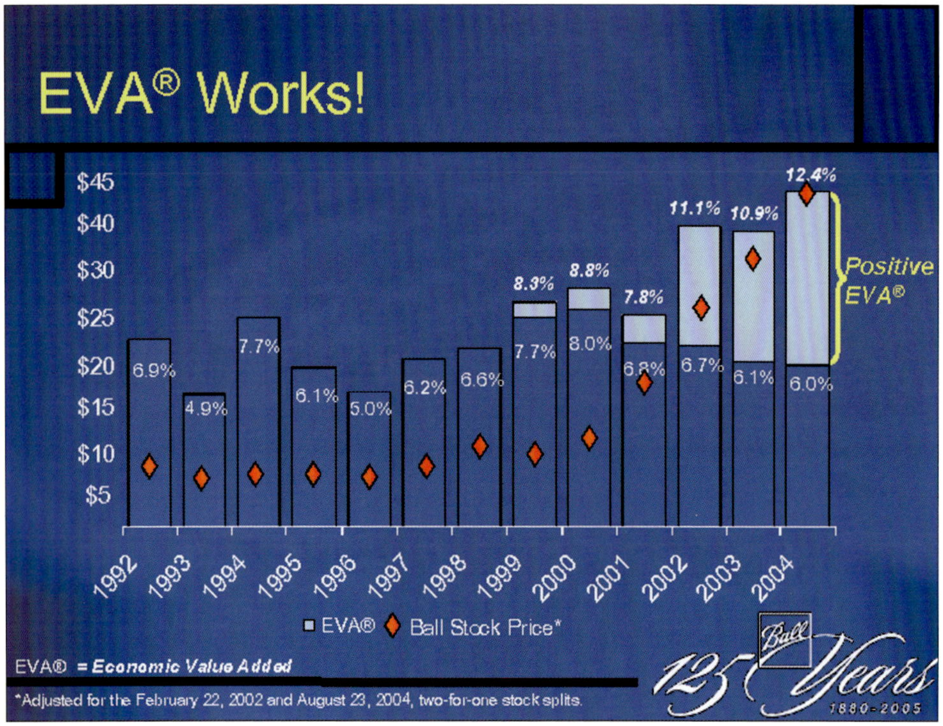

Ball uses a statistic called EVA® (Economic Value Added) to evaluate its performance. The company is considered to be creating economic value when its after-tax returns exceed its cost of capital. The graph above, plotting Ball's EVA between 1992 and 2004, shows a close correlation between positive EVA and stock-price improvement.

approximately $25 million in cash and stock. Not finding a buyer for Ball Aerospace was the best thing that could have happened. Ball Aerospace has improved its results significantly and is today a healthy, vital part of Ball Corporation, worth many times what that suitor offered in 1994.

In 1995, the company began manufacturing PET [polyethylene terephthalate] plastic bottles, but the new business struggled for several years.

Each of these initiatives is discussed in greater detail in the chapters that follow.

Financial Discipline and Difficult Decisions

Ringoen retired in 1991 and was succeeded as president and chief executive officer by Delmont A. Davis Jr., who, like Ringoen, had a technical background. Davis was born near Hillside, Colorado, in a two-room log cabin built by his father, a homesteader from Kansas. He graduated from the University of Colorado in 1959 with a degree in civil engineering and worked at the Federal Aviation Administration before joining Martin Marietta and then General Electric as a project manager in the Apollo moon-landing program. He was hired by Ball Aerospace in 1969, transferring several months later to Ball's metal container group as director of engineering and development. He subsequently was the group's vice president of operations and then group vice president, becoming executive vice president of the packaging products group in 1987.

In 1989, Davis was named president and chief operating officer of Ball, succeeding Ringoen as CEO in 1991. Longtime director Alvin Owsley, a grandson of Frank Ball and a retired senior partner in the Houston law firm Baker & Botts, became chairman of the board. Ringoen continued on the board.

The year after Davis became CEO, Ball took the crucial step of adopting EVA® (Economic Value Added), an acronym coined by management consulting firm Stern Stewart & Co. Dave Hoover, then vice president and treasurer, was—and still is—a champion of this management tool for assessing how a company's strategies and investment decisions create or destroy value.

Using EVA, companies analyze their after-tax returns on capital (all the money tied up in plant, equipment, inventories, receivables and other items) against the cost of that capital. Economic value is created when after-tax returns exceed the costs of the capital invested in that business. This is similar to when an individual borrows $1,000 to invest in the stock market, pays interest of $100 on the borrowed money and earns a stock-market profit of $250. The net gain (value added) is $150.

In 1993, *Fortune* magazine called EVA "the real key to creating wealth" and noted that the concept was "catching on quick" among corporations. Ball was one of its pioneers, joining Quaker Oats, Coca-Cola, Briggs & Stratton and other early adopters.

Although EVA is now an essential tool for Ball, it was not a magic bullet that immediately improved the company's results to satisfactory levels. Hoover says, "We adopted EVA in 1992 and it drove the performance of our operations. But we did not generate returns in excess of our cost of capital until 1999."

Nonetheless, EVA provided Ball with the financial discipline to make difficult decisions and reshape the company into a more focused, successful enterprise. One of these decisions was the sale of its commercial glass operations. "Using EVA, we saw that the glass business was not generating adequate returns and we would be better off deploying our capital in other businesses where the returns were better," says Ray Seabrook, senior vice president and chief financial officer.

In 1993, Ball spun off seven of its smaller businesses into an independent new company, Alltrista Corporation. These operations included canning jars and Ball-branded products, below, industrial plastics, injection-molded plastic products, plastic packaging, metal decorating and services, quality-control inspection systems and zinc products. Former Alltrista businesses are now part of Jarden Corporation.

Spinning Off Alltrista

EVA also helped guide another of the major decisions made by Ball in the 1990s: the formation of Alltrista Corporation and the distribution of its shares to Ball stockholders in 1993.

Ball's senior management team felt the company had too diverse a mixture of businesses. Thus, seven of Ball's smaller businesses—consumer products, including home canning jars; zinc products; industrial plastics; injection-molded plastic products; plastic packaging; metal decorating and services and quality-control inspection systems—were combined into an independent company, Alltrista, which was spun off to Ball shareholders and led initially by two former Ball executives, William L. Peterson and Thomas B. Clark. Alltrista is known today as Jarden Corporation and is listed on the New York Stock Exchange.

Each of the businesses had potential in its own right but added little to a large company such as Ball. In fact, the seven businesses had combined annual sales of $193 million, less than 10 percent of Ball's total revenues, and consumed a disproportionate amount of management time and capital investment.

The *Indianapolis Business Journal* reported in April 1993, "The move is designed to allow the century-old Fortune 500 company [Ball] to focus its efforts on its core operations: beverage and food container manufacturing, and aerospace systems and services. The move will also allow Alltrista the freedom to nurture the smaller companies it inherits."

Ann Scott, Ball's director of investor relations and analysis, says the spin-off was "a big change for the company, its employees and the community," not only because some operations and their employees left Ball to join Alltrista but also because the spin-off signaled Ball's seriousness about becoming more disciplined and profitable.

Most striking was Ball's decision to jettison its century-old home canning jar business. Although some employees and consumers wondered

why the company would sever this link with its past, canning jars were now a tiny part of Ball and generated minimal profits. Ball brand canning jars are still marketed today—only now by Jarden under a license to use the Ball name and historic script logo on home canning products.

Management Succession Challenge

In May 1994, Davis resigned as CEO after three years in the position. His resignation triggered a mini-crisis: Ball did not have a successor in place and searched nearly a year before it found one.

The board named George Sissel—senior vice president of corporate affairs, corporate secretary and general counsel—acting president, while a three-member committee of the board looked for a new CEO. Aided by an executive-search firm, the board considered 32 candidates before narrowing the field to six finalists consisting of internal as well as external candidates. Completing the process in late 1994, the board chose one of the six candidates—an executive of an industrial firm—and offered him the job.

Then the unthinkable happened. After initially accepting the offer, the candidate changed his mind and turned it down. "Ball Rejected by CEO Pick," a headline in the *Muncie Star* reported on January 17, 1995. The *Muncie Evening Press* quoted Ball Chairman Alvin Owsley as saying, "We had a compensation package. It was all peaches and cream. Then he told the search committee his wife would not move."

Left in the lurch, Ball eventually terminated its search and removed the word "acting" from George Sissel's title, making him CEO in April 1995. Owsley said Sissel, 58, "was doing a terrific job" and had earned the promotion.

George A. Sissel was born in Chicago and attended high school in Colorado. He graduated from the University of Colorado in 1958 with an engineering degree, served as a navy officer and graduated from the University of Minnesota law school in 1966. He worked at a Denver law firm before joining Ball in 1970, becoming assistant corporate secretary and general counsel eight years later. By 1987, he had worked his way to senior vice president.

When Sissel became CEO in 1995, Ball's management team laid out a five-part strategy: maintain leadership as a low-cost, high-quality North American beverage can manufacturer; position North American rigid packaging for significant earnings and positive cash flow; establish Ball in PET plastic containers; expand internationally in packaging; and capitalize on the world-class capabilities of Ball Aerospace. Ball has since made excellent progress toward each of these objectives. Faced with new challenges, the company once again evolved to meet the needs of a changing world.

"The [spin-off] is designed to allow the century-old Fortune 500 company [Ball] to focus its efforts on its core operations: beverage and food container manufacturing, and aerospace systems and services."

In 1998, Ball moved its corporate headquarters from Muncie, Indiana, to a building, left, in Broomfield, Colorado, that had previously housed parts of its aerospace business. The company had fewer than 200 employees in Muncie compared with 3,000 at its packaging and aerospace operations in Colorado, contributing to the move.

THE NEWSPAPER FOR MUNCIE AND EAST CENTRAL INDIANA

The Star Press

THURSDAY, February 5, 1998

Single copy 50¢

"Ball has been working with this kind of question for the past 5 years. Our business has changed."

George Sissel
Ball chairman and CEO

Kyle Evens / The Star Press

Ball Rolls West

Headquarters move to Colorado ends century-old link to city of Muncie

By DAVID PENTICUFF
The Star Press

MUNCIE — Ball Corp., the community's signature business for more than a century, will leave Muncie by the end of this year.

The dwindling presence of Ball in Muncie and Indiana led to the decision, which was announced Wednesday by George Sissel, chairman and chief executive officer.

Edmund Ball, former head of the company and son of one of the company's founders, and also with John Fisher, a Ball family member and former Ball Corp. CEO.

"It is a sad day for them personally," Sissel said. "It is a day of realization for them. They felt, sooner or later, this would happen."

It was not in the best interests of Ball to stay, Sissel said

The Move from Muncie

In 1998, Sissel received high marks for his handling of a potentially contentious issue: the company's decision to move its headquarters to Broomfield, Colorado, after having called Muncie, Indiana, home for more than a century.

Ball had fewer than 200 employees in Muncie as of 1998 and had sold or spun off all of its production facilities there, leaving only the corporate headquarters. After announcing the relocation on February 4, Sissel told a packed press conference in Muncie that the move was a "terribly painful process to think about because of

Four current and past senior executives—David Hoover, John Fisher, Ed Ball and George Sissel, left to right—came together in 1998 to thank the people of Muncie.

all we have invested" in the city. He also said, "We've had a great 111 years here. This community gets under your skin and stays there." Sissel emphasized that it made economic and business sense to relocate to Colorado, where Ball had 3,000 employees in its packaging and aerospace operations.

Initial public reaction in Muncie was mostly negative. The second question asked at the press conference was whether the Ball brothers "were rolling over in their graves." That evening, some callers to local radio talk shows said they could not believe that Ball, its name virtually synonymous with that of their city, planned to leave.

Sissel worked hard to calm frayed nerves. He met with community leaders and others to explain why and how Ball had reached its decision. In time, the local media rallied to Ball's side, thanking the company for its civic contributions over the years—some of which it promised to continue after leaving. The *Muncie Star Press* headlined a June 1998 editorial about the relocation "A Benevolent Exit."

Some employees in Muncie were nervous about the move. Many had never lived anywhere except Indiana and did not know what life would be like in the Rocky Mountain state. Doug Bradford, vice president and controller, says Ball went out of its way to work with employees and treat them fairly. "The company created a 'soft landing' in Colorado by providing preview trips and offering moving assistance and arranging for a real estate agent to help employees find new homes," Bradford said. Most of the 70 employees who moved found they enjoyed life in Colorado and made the adjustment quickly.

Relocating to Colorado not only reduced costs by consolidating jobs, centralizing offices and eliminating frequent travel between the two states but also provided intangible benefits, such as erasing decades of Colorado-versus-Muncie corporate culture. Colorado also proved to be a drawing card for new employees such as John Hayes, hired in 1999 as senior director, corporate planning development, and Scott Morrison, vice president and treasurer. Both executives joined the company after the move. In 2005 Hayes was named executive vice president of Ball Packaging Europe.

1998-2005: Growth and Value Creation

By 1998, despite all the changes made by the company during the prior decade, Hoover (then executive vice president and chief financial officer) and others recognized that dramatic steps still needed to be taken to improve Ball's performance. "We realized we weren't the biggest or the best in anything," he says.

Thus Ball jumped at the opportunity to acquire the North American beverage container business of Reynolds Metals Company in 1998. With this acquisition, Ball went from being North America's fourth largest beverage can manufacturer to its largest.

Two years later, in 2000, Hoover was named CEO, succeeding Sissel, on January 1, 2001. Known for his communication skills and ability to articulate the company's strategies and goals, he had been a chief architect of the Reynolds acquisition and had run each of the company's businesses, packaging and aerospace.

On becoming CEO, Hoover was asked about his expectations of employees. He replied, "Everything ought to be crystal clear to people in terms of what I'm looking for out of this company. We are competing for capital. We have to perform. We have to create value. And the minute that we don't do that, we don't stay focused on achieving profitable growth and earnings and making a good return on the investment that shareholders entrust with us, we're not the Ball Corporation that I know."

Hoover wasted little time putting his words into action. In 2001, Ball formed a 50-50 joint venture with Coors Brewing Company to make cans for its large Golden, Colorado, brewery. In 2002, Hoover and Hayes spearheaded the company's acquisition of Schmalbach-Lubeca GmbH, Europe's second largest beverage can producer. By 2004, Ball also had added to its steel food can business by first forming a joint venture to own a ConAgra Foods, Inc., plant in Oakdale, California, and later acquiring the plant—solidifying its position as North America's second largest manufacturer of food cans.

With these and other transactions, Ball became larger and more highly focused than ever. In fact, the period since 1998 has witnessed a dramatic recovery from the challenges of 1987-1997. Sales rose from $2.99 billion in 1998 to $5.44 billion in 2004, an 81 percent increase, while split-adjusted earnings per share advanced from 48 cents per share to $2.60 in the same period, up more than fourfold.

Each of the company's business segments—North American packaging, international packaging, and aerospace and technologies—posted record earnings in 2004.

CHAPTER FIVE:

Good Things Come in Metal Packages: Two-Piece Beverage Cans

The aluminum beverage can boasts an unusual history. *Progressive Grocer* magazine once described it as one of the "most influential products of the 20th century," putting it on a list with innovations such as homogenized milk, the safety razor and the ice cream cone.

The steel beverage can made its debut in 1935 but had limitations because of its weight and cost. In response, Adolph Coors Company developed the first practical aluminum can. The new package quickly gained favor with beverage companies and consumers. Hawaii Brewing Company pioneered its use in 1958 and Coors followed suit the next year. Coke and Pepsi started selling soft drinks in aluminum cans in 1967.

In 1979, the year in which the Can Manufacturers Institute began keeping statistics, 34 billion aluminum beer and soft drink cans were produced in the United States. By 1989, production had reached 83 billion cans a year. Today the aluminum can is one of the most ubiquitous packages in the world. In the United States, each person, on average, purchases one beverage in an aluminum can every day of the year.

There are many reasons for this humble product's great success. The aluminum beverage can is not only lightweight, convenient, durable and stackable, but also recyclable. What is more, it chills faster than other packages, feels colder than plastic, doesn't break like glass and has a long shelf life. In addition, its surface can be printed with eye-catching graphics that help sell beverages.

Ball is North America's largest manufacturer of aluminum beverage cans—supplying them in a variety of sizes for Budweiser, Coca-Cola, Coors, Heineken, Miller, Molson, Pepsi-Cola and other well-known brands.

Ball has a growing position in the custom beverage can business, making unique sizes and features that help customers differentiate their products. Go Fast! Energy Drink promotes its unique 11.9-ounce serving size as "perfect." Other Ball customers use colored tabs to coordinate with graphics and draw consumers' eyes.

The aluminum can continues to evolve. Recent years have seen new sizes, such as the 8.4-ounce "slim" can for energy drinks; new designs, including reclosable screw-top cans; new uses, such as for wine; and technological advances that allow cans to be manufactured better, faster and less expensively than ever.

Ball is involved in every one of these developments. After all, no product is more important to Ball than the aluminum beverage can. About two-thirds of the company's worldwide revenues each year are derived from the manufacture and sale of this modern packaging marvel. In the huge North American market, one of every three aluminum cans made each year is produced by Ball.

Worldwide, Ball makes more than 50 billion recyclable aluminum beverage can bodies and about as many can ends every year. Some companies that purchase cans from Ball purchase the ends from other sources, and vice versa. The bodies and ends are joined on beverage companies' production lines after the cans have been filled. Because the contents are carbonated, aluminum cans are engineered to withstand internal pressure of more than 90 pounds per square inch, three times the pressure in a typical automobile tire.

The economics of manufacturing, transporting and storing can bodies are quite different from those of can ends. The bodies are expensive to ship long distances because truckloads of empty cans consist largely of air. As a result, Ball operates 19 aluminum can body factories across North America, close to customers' beverage filling lines. Ends, by contrast, can be packed tightly and shipped economically over long distances. Consequently, Ball's end manufacturing in North America is consolidated in only a few factories for maximum production efficiency.

The Acquisition of Jeffco Manufacturing

Ball's growth in aluminum beverage cans is a classic American success story. The company entered the business relatively early, grew its capacity steadily for three

The Best Recyclable Package

Aluminum cans are the world's most recycled consumer beverage containers, making them uniquely valuable. Among their advantages:

- Aluminum is recyclable endlessly. It can be recovered, reprocessed and reused over and over without degradation.
- Recycling reduces landfill and saves energy. The processing of recycled aluminum requires only 5 percent of the energy needed to manufacture aluminum from ore. Recycling a single can saves enough energy to power a television for three hours.
- The packaging industry's down-weighting of cans has resulted in the conservation of metal. In 1980, a 12-ounce aluminum beverage can weighed about 0.7 ounces, equal to 24 cans per pound of aluminum. Today's cans weigh less than 0.5 ounces, about 33 cans per pound.

The aluminum industry has created a nationwide aluminum can recycling network unrivaled by any other packaging material. In 2004, about 51.5 billion aluminum cans, or just more than half of those sold in North America, were recycled.

Ball, together with other corporations, industry groups and government, is seeking to increase recycling rates with an ultimate goal that every can be recycled. Some Ball facilities are promoting recycling in unusual ways. Employees at Monticello, Indiana, worked with the local community to raise money for a public skate park by collecting and recycling cans. Ross Rittberg, Monticello plant manager, says the effort provided funds for a worthy cause while increasing recycling in the county by more than 100 percent. The Fort Worth, Texas, plant participated in the America Recycles Aluminum Day Can Challenge, collecting cans and donating proceeds to a local cancer care agency. In Colorado, Ball is partnering with Coors Brewing Company to fund community recycling drop-off centers.

"Recycling is very important to us," Jim Fisher, Ball's vice president, packaging industry affairs, packaging operations, states. "It benefits the environment and is a win for everybody—the public, our customers and Ball."

Harry "Hank" Schroeder, environmental manager at the Findlay, Ohio, metal food and beverage container plant, developed a recycling program that reduced the plant's waste material by more than 40 percent and facilitated the recycling of more than 2,000 tons of materials a year.

High-speed production lines at Ball's plants are able to manufacture 2,000 or more can bodies per minute—making the job of Chief Quality Assurance Inspector Cristy Wise and her colleagues especially important. At right, Wise inspects a batch of cans at the plant in Findlay, Ohio. Ball entered the beverage can business in 1969 with its acquisition of Jeffco Manufacturing Company, which at the time presented lighters to employees celebrating their first year of service.

decades and then leaped to the top of the industry in 1998 when it successfully completed an acquisition that skeptics doubted would ever happen. In the process, it has served the needs of beverage companies and consumers by making quality cans ever more efficiently, helping keep packaging costs down despite economic inflation. The aluminum can typically represents less than 10 cents of the total cost of a canned beverage.

Ball made its first beverage can in 1969 when it acquired Jeffco Manufacturing Company, one of the early innovators. Jeffco's facility in Golden, Colorado, supplied can bodies and ends to Coors and was in the process of converting production from steel cans to the newly developed two-piece aluminum can, which was less expensive to make than the original three-piece can because it used less metal. Ball gained a competitive advantage, *Barron's* wrote in 1983, by manufacturing two-piece cans "ahead of competitors who were slow to convert from three-piece cans." Today, three-piece beverage cans are no longer manufactured in North America; all can production is two-piece.

Jeffco had been founded in 1961 by Karl Nelson, a Nebraska businessman who also owned Morton House Kitchens, a canned-foods company that purchased metal decorating services from Ball. "One of our salesmen was very close to Karl and learned that he wanted to sell his company in Golden," retired Ball CEO John Fisher says. "We talked with Karl and studied the company carefully. We also studied the industry's technology." Ball liked the opportunities it saw in the beverage can market and believed, in particular, that it could apply its metallurgical know-how, developed in its zinc products business, to the manufacture of aluminum cans. Ball paid for Jeffco by issuing Ball stock, and Nelson joined the company's board of directors.

The Golden plant continues today as a vital part of Ball. Repeatedly expanded and modernized, it now annually manufactures as many as 2.2 billion beverage can bodies for soft drink companies in the Rocky Mountain region and up to nearly 5 billion can ends for soft drink and beer companies throughout the western United States. "It is one of the most productive and innovative beverage can plants in the world," according to the *Denver Post.*

Early Growth in the Beverage Can Business

Ball's beverage can business was led in its early years by a team of young, energetic managers with backgrounds in engineering and aerospace. They included Delmont Davis, future Ball president and chief executive officer; David DiLoreto, future group vice president, packaging products; David Sheldon, executive vice president, packaging operations; and William Lincoln, future executive vice president, metal container operations. DiLoreto says, "From day one, we defined our business as being manufacturing-driven, and we focused on achieving high quality and low costs." He also says Ball asked its customers to deal directly with plant managers on quality and production issues rather than communicate exclusively through the company's sales staff, as most competitors did. "Direct communication between our plants and customers was vital to our success," he says. "It strengthened our customer relationships and helped ensure that problems got resolved quickly."

In addition to manufacturing cans, Jeffco provided can making equipment and consulting services to others. "We encouraged them to go ahead with their technology and consulting business," Fisher says. One year after being acquired by Ball, Jeffco was retained by Joseph Schlitz Brewing Company to engineer, design and assist in the construction of a can factory for Schlitz in Milwaukee.

Ball also used Jeffco's technology to build new Ball-owned plants. Production at each of these facilities was dedicated to a particular customer. Even though Ball was a newcomer to the business, it was able to win supply agreements with some of the nation's top beverage firms because of the reputation it had for manufacturing excellence and customer service.

Ball made its first cans for Anheuser-Busch in 1971 when it built a $12 million factory in Williamsburg, Virginia, to serve that company's newly opened Williamsburg brewery. Because Anheuser-Busch is a demanding customer with high quality-control standards, "we realized that much of our company's success in the metal container industry depended on the performance of this plant and its workforce," Fisher recalls. Unfortunately, the plant had startup problems. With so much at stake, one plant manager was let go and then another. Ball subsequently hired George Matsik from Continental Can to become the facility's production manager. He quickly got costs under control, improved quality and turned the factory into one of the best in the industry. Pleased with this performance, Anheuser-Busch asked Ball to supply cans for a brewery being built in northern California, resulting in the opening of Ball's Fairfield, California, plant in 1976. In 1981, Ball erected a plant in Saratoga Springs, New York, to supply other Anheuser-Busch breweries.

Speaking of Williamsburg, Fairfield and Saratoga Springs, Matsik quips, "Those three plants kind of gave us the Good Housekeeping Seal of Approval in the beverage can industry." As of 2005, Ball supplies cans to Anheuser-Busch from three factories—Williamsburg, Saratoga Springs and Findlay, Ohio. All three have

Aluminum cans are one of the most ubiquitous packaging products. They are lightweight, convenient and recyclable and can be stacked efficiently in a store display or home refrigerator. Moreover, they chill faster than other packages, don't break and can be printed with high-quality graphics.

Ball's Cal Winslow leans on a coil of aluminum alloy used in the manufacture of can ends, which are attached to can bodies on customers' filling lines. As of mid-2005, Winslow, end line chief maintainer at the company's Fairfield, California, beverage container plant when the photo above was taken in 1981, hadn't missed a day of work in 17 years. He is currently a rebuild/maintenance chief at the same plant.

earned "Certified Supplier" status with Anheuser-Busch, based on providing the highest level of service and quality. In fact, Ball is one of only about a dozen Anheuser-Busch suppliers to have received this elite award since its inception in 1993.

Meanwhile, Stroh was looking for a supplier to make cans for its brewery in Detroit. It chose Ball, which in turn picked Findlay, about 100 miles south of Detroit on Interstate 75, as the site of a new can factory. "Our people came down I-75 until they found a place they liked," Tom Martin, plant manager at Findlay, says. All the early Ball beverage can plants were located in smaller and mid-sized communities such as Findlay, which has a population of about 35,000. "This was very deliberate," George Sissel says. "John Fisher, Dick Ringoen and Del Davis wanted all our factories to be located in attractive places to live. They sought out smaller communities that would appreciate having a Ball plant and where we could hire competent hourly and salaried personnel."

The Findlay plant was completed in 1972. In 1980, it moved from producing cans for Stroh to supplying Coca-Cola and Pepsi-Cola, adding a steel food can line in 1996. Today Findlay is one of the largest Ball manufacturing facilities in the world, producing 1.8 billion aluminum beverage cans, 12 billion can ends and 700 million steel food cans a year.

Some of the engineers who joined the company's beverage can business in its early years came from Ball's aerospace business, giving the company's can business tremendous technical skill that translated into a key competitive advantage. "There was, and still is, an appreciation of the importance of technology in our packaging operations, where engineering plays a vital role," observes John Friedery, senior vice president and chief operating officer, North American packaging operations.

Seeking To Become the Industry Leader

Although Ball quickly became known as a quality supplier, it was much smaller than industry giants National Can and Continental Can. Harold Sohn—who joined Ball in 1977 and today is vice president, corporate relations—recalls attending a Can Manufacturers Institute (CMI) meeting at which someone from National Can teased him good-naturedly as the men walked through a buffet line. CMI dues were based in part on a manufacturer's size in the industry. "The guy from National said, 'We bought this here,' and he looked at the meat," Sohn recalls. "And then he looked at another main dish and said, 'You guys from Continental bought that.' And he looked at the little relish tray and said, 'You guys from Ball bought the relish.' He was right in his analogy. We were the little guy in the industry."

Growing its beverage can business became a priority. To spur that growth, the company invested more money in its can making operations than in any other product line in the 1970s and 1980s. As a result, by the late 1980s, Ball operated nine beverage can factories from coast to coast and was the fourth largest aluminum can manufacturer in North America.

Seeking to get bigger and more competitive, as well as to expand globally, the company agreed in 1990 to purchase Continental Can Europe (CCE), a top overseas can maker, for $1 billion, an enormous amount for Ball at the time. *Barron's*, noting that the transaction would nearly triple Ball's annual sales, said the company was "going into a Jonah-swallows-the-whale act."

But the deal was stymied by a sudden shift in financial markets: banks, seeing a global economic recession on the horizon, cut back sharply on their lending to ride out the economic storm. Unable to raise money to finance the planned purchase, Ball was forced to abandon the deal three months after it was announced. In a joint press release, Ball and Peter Kiewit Sons' Inc. (which owned CCE) said, "Unfortunately, due in major part to the recent turmoil in world financial markets, and in particular the U.S. banking community, Ball has not yet arranged financing, and the certainty of closing the transaction on the agreed-upon terms and time frame has been compromised."

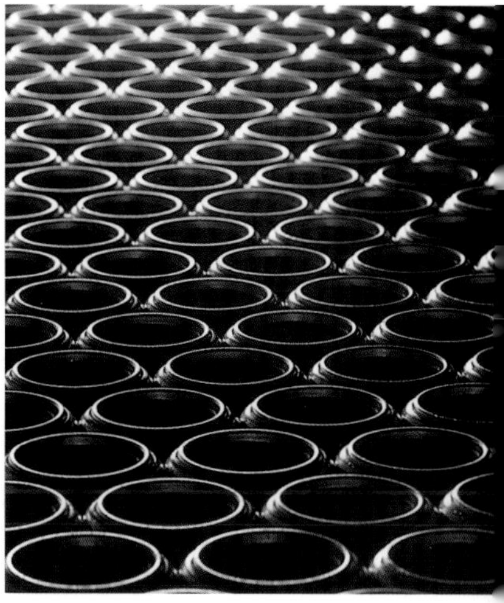

Beverage cans await shipment from the warehouse at the company's plant in Williamsburg, Virginia, opposite. Completed in 1972, the Williamsburg facility was Ball's second beverage can plant. Today, Ball operates nearly 40 wholly owned and joint-venture beverage can and end plants worldwide. Beverage cans are Ball's largest business, accounting for about two-thirds of revenues in 2004.

Bill Dumont, a front end maintainer at the time this photo was taken and now an operator at Ball's beverage can plant in Saratoga Springs, New York, performs a quality check. "Whenever we can get customers into our facility, we beat down the doors to do it," says Steve DiLoreto, manager of the plant, which serves customers in the soft drink and beer industries. "There's nothing that sells Ball better than for someone to see our facilities and meet our people."

CCE was subsequently acquired by Viag AG of Germany. "We gave it the good old college try," Hoover says, "but at the end of the day we were glad that Viag came along. We would have owed them [Kiewit] a breakup fee, but when Viag purchased CCE that pretty much let us off the hook." He adds, "Those were very dark days. We tried and failed to do something."

Nonetheless, the story had a happy ending. Twelve years later, Ball acquired CCE's successor, Schmalbach-Lubeca, and in this way bought most of the same European beverage can plants it had sought to obtain in the CCE transaction.

Acquiring Reynolds' Can Business

By the mid-1990s, Ball faced an increasingly critical need to get bigger in the beverage can business and achieve economies of scale or risk falling hopelessly behind its competitors. At that time, beverage can demand in the United States was beginning to peak. Demand reached an all-time high of 103 billion cans in 1994 and has been flat to slightly lower ever since. Among all the company's businesses, beverage cans were generating the highest after-tax returns on capital. Yet, to maintain and improve those returns in a consolidating market, Ball needed to become larger and more efficient.

"We looked around and said, 'We aren't a leader in anything,'" Hoover recalls. "One of the things we did best was make beer and soft drink cans, but we were tied for fourth in size with Reynolds. We basically said we had to get bigger to survive."

Could Ball, a mid-sized company with sales of $2.4 billion in 1997, pull off a major acquisition?

There were many skeptics. One investment firm wrote in 1993 that Ball, having missed out on purchasing Continental Can Europe, "has no obvious opportunities for acquisitions," adding that "the most plausible target in the can business—Van Dorn—has just been acquired by Crown Cork & Seal." The investment firm said flatly that "there are no good candidates left" to be acquired in the beverage can industry.

Ball proved the skeptics wrong. Determined to grow, the company found an acquisition that literally changed its future. As of 1998, Ball and Reynolds Metals Company were the two smallest players in a five-company North American beverage can industry. Each held a market share of about 17 percent. Ball acquired Reynolds' can business to surpass its larger rivals in one giant step and become the number-one beverage can maker in North America, a position it holds to this day.

Ball proved the skeptics wrong. Determined to grow, the company found an acquisition that literally changed its future.

New senior management sparked the move. In 1997, Dave Hoover was promoted to vice chairman while continuing as chief financial officer, and George Matsik became president and chief operating officer, packaging. George Sissel, who was chairman and chief executive officer, gave his two newly promoted associates leeway to run the business. They rose to the challenge.

Hoover and Matsik sat down to discuss Ball's future. "We looked at about ten scenarios, including buying or merging with various other companies," according to Matsik, who is now retired. As a result of the goals outlined at their meeting, Ball explored potential acquisitions or mergers. "We could have done something, I believe, with a number of other companies," Hoover says.

Ultimately, Hoover and Matsik decided to focus on acquiring Reynolds' can business, which offered the best fit. Reynolds operated 16 can plants in 12 states and Puerto Rico, while Ball had nine in North America. Some of Reynolds' facilities were located in regions that

BELVAC
BELVAC
BELVAC PRODUCTION MACHINERY
BELVAC

Ball did not serve, such as the Pacific Northwest, the Carolinas, Hawaii and Puerto Rico. Purchasing Reynolds' can division would give Ball its first presence in those markets. The economics of acquiring Reynolds' can business, including the projected cost savings and the ability to expand into new markets, were overwhelmingly positive.

Like Ball, Reynolds had a new management team that was looking to make a deal. Reynolds' new chairman and CEO, Jeremiah "Jerry" Sheehan, had inherited a company that was not performing well. He began selling off assets, hinting in October 1996 for the first time that he might sell the company's aluminum can division—a statement that piqued the interest of Ball. Ten months later, in a speech to an investment group, he went further by stating that Reynolds planned to "optimize" the value of its can business "one way or another"—perhaps by selling it or spinning it off to Reynolds shareholders.

There were many parallels between Ball and

The Monticello, Indiana, beverage can plant, opposite and above, is one of the facilities that joined Ball in its 1998 acquisition of the beverage can business of Reynolds Metals Company. The Reynolds acquisition not only expanded Ball's manufacturing capacity, but also brought new customers such as Miller Brewing Company, now SABMiller.

Reynolds. Both companies started as family-owned businesses (Reynolds was founded by Richard S. Reynolds Sr. in 1919), and both owed their early success to a popular consumer product: Ball with its promotion of the home canning jar and Reynolds with its invention of aluminum foil. Moreover, both changed as their markets shifted. Reynolds began producing aluminum metal after World War II and started making aluminum cans in 1963 as a way to sell more of that metal. However, the company failed to invest adequately in the modernization of its can plants, with the result that they became less and less competitive. Selling the plants would allow Reynolds to refocus on its core business of aluminum production.

In mid-1997, Hoover, Matsik and Sissel traveled to New York to meet with Sheehan and discuss the possibility of buying the Reynolds facilities. At that meeting, Sheehan quoted an asking price of $1.2 billion for his company's domestic and South American beverage can business, a price all three Ball executives found extremely steep. Undeterred by the opening price, Hoover and Matsik kept bargaining for nearly a year until Reynolds finally quoted a price they liked.

Following that initial meeting in New York, the two executives traveled regularly to Reynolds' headquarters in Richmond, Virginia, to negotiate with members of that company's senior management team. Hoover says, "We just sort of 'smoked the long cigar' and kept meeting and saying, 'It's too much, too much.'"

Unrelated to the negotiations, Ball announced in February 1998 that it was moving its corporate headquarters from Indiana to Colorado. Moreover, throughout the period of the negotiations, Ball was dealing with the financial problems of its beverage can business in China. "It was a crazy time," says Jim Fisher, vice president, packaging industry affairs, packaging operations. "So many things were happening at once."

In addition, the negotiations were complicated by several factors unrelated to price. One issue was whether Reynolds' 35 percent interest in a Brazilian aluminum can business called Latasa would be included in the deal. "That was a big sticking point," Hoover says. When Ball and Reynolds first announced publicly in April 1998 that they had reached a deal, they said Latasa was included. *American Metal Market*, a trade journal, speculated that Ball wanted to combine its existing can making operations in Brazil with Latasa to create a "huge economy-of-scale advantage over its competition" in South America. Two months later, however, Ball and Reynolds announced that Latasa would be excluded from the sale, prompting *American Metal Market* to report that the companies "couldn't agree on a price for Latasa." Recalling these events, Hoover says Ball actually preferred from the beginning that Latasa be excluded. "We didn't want to buy it but they wanted to include it," he says. "We were finally able to carve it out from the purchase." As a result, Ball acquired only Reynolds' domestic can business.

Ball worked hard to improve the performance of the acquired Reynolds plants and provide superior products and service to Miller and other customers. Those efforts were recognized in 2001 when Ball received Miller's coveted Supplier of the Year award.

Another complication was Reynolds' planned sale of its rolling mill in Muscle Shoals, Alabama, to a leveraged buyout (LBO) group. Reynolds did not want to sell its beverage can plants unless it could sell the mill, which supplied aluminum sheet to the can industry. However, the mill buyout group needed to have a signed agreement that the mill would continue to supply aluminum to the Reynolds can plants before lenders would provide financing for the buyout. Thus, Reynolds' sale of the Alabama mill and the sale of its 16 can manufacturing plants had to be closely coordinated even though the buyers were different. Ball, as the prospective acquirer of the can plants, negotiated with the LBO group to purchase aluminum from the mill, ultimately signing a contract that enabled the buyout group to get its financing.

Closing the Deal

When Reynolds first hinted that its aluminum can business might be for sale, several interested bidders—not just Ball—lined up. One by one, however, they dropped out until only Ball was left.

Reynolds' asking price eventually got below $850 million including Latasa, at which point Sheehan—who had left the day-to-day negotiations to other members of his management team—returned to the table and made a final compromise that sealed the deal. Ball ended up paying $745 million excluding Latasa.

On April 23, 1998, Ball announced that it had reached a definitive agreement to acquire Reynolds' beverage can manufacturing business in North America. Although the news was exciting for Ball and its people, it laid an egg on Wall Street. Many investment analysts questioned whether Ball was capable of managing such a large acquisition. Reflecting this concern, Ball's share price dropped 3 percent the day the transaction was disclosed and continued to decline for three weeks before rallying. Hoover and Matsik, confident that the deal would benefit Ball and its shareholders, ignored the market's reaction and pushed ahead.

However, the company still needed to resolve some financing issues in China before it could borrow money to buy the can plants in the United States. With the board of directors' approval, the company invested additional equity capital in its Hong Kong subsidiary, meeting the requirements of the subsidiary's locally financed debt. Ball was then able to arrange the financing it needed to buy the Reynolds plants.

On August 10, 1998, Ball completed the acquisition, ending a nearly year-long odyssey. Asked if he was thrilled when the deal closed, Hoover replied, "Absolutely. It was very much a defining moment in our company's history."

Reynolds Acquisition Comes Up a Winner

Having acquired the Reynolds plants, Ball went about the process of integrating them into the Ball supply chain and maximizing their value. Ball people had worked hard throughout the negotiations to understand the strengths and weaknesses of every one of the Reynolds facilities. During the negotiations, Ball had sent teams of employees to each plant to study its operations in depth and identify opportunities for improvement. "By the time we closed the deal, we knew more about the plants individually than Reynolds did," according to Charles Baker, who was then senior director, business development, and is today vice president, general counsel and assistant corporate secretary. As a result, Ball was able to move quickly to reduce costs and invest in manufacturing improvements.

The deal offered significant logistical savings. Now that it had more beverage can plants in different locations across the United States, Ball was able to serve some of its customers from plants that were nearer to them, reducing freight costs. Ball's plant in Findlay, Ohio, for instance, served some customers two states west in Illinois, while Reynolds' plant in Monticello, Indiana, served some customers one state east in Ohio. "Our drivers could wave to each other as they passed on

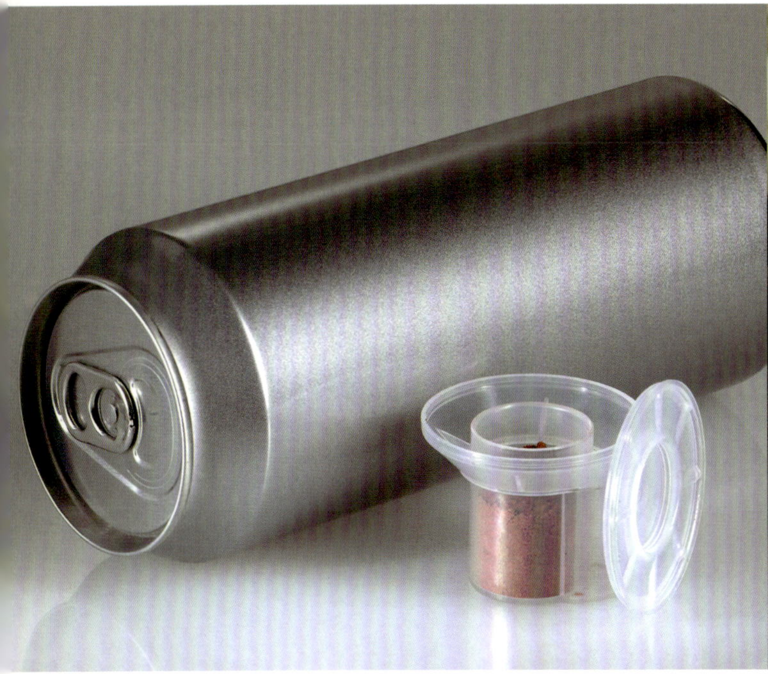

Ball produces an innovative, sleek beverage can, above right, that is a recyclable and practically spill-free alternative to the juice box. Ball's wine can, center, is easy to hold and ideal for poolsides. The wedge, left, is a small plastic container, suspended in a can, that stores dry ingredients such as vitamins that would quickly degrade in liquid form. It injects its contents into a beverage as soon as the top is popped.

the interstate," Leon Midgett, retired executive vice president and chief operating officer, packaging, remarks. Following the acquisition, new customers in Illinois were served from Monticello and new customers in Ohio from Findlay. "We kick and scream and fight and struggle to get even five cents per thousand cans out of our costs," Midgett says. "The Reynolds acquisition allowed us to reduce our costs by more than that through shipping efficiencies alone."

In addition, the two companies' facilities overlapped in some regions, such as central Florida. As a result, Ball was able to reduce costs and operate more efficiently by later closing some plants that were close to other plants. Ball needed to achieve $70 million in total savings from the acquisition for it to have a positive impact on earnings. The actual savings turned out to be considerably greater.

In addition to bringing new facilities and talented people to Ball, the acquisition added new products and new customers. Prior to the deal, Ball primarily made 12-ounce and 16-ounce cans, the traditional bread and butter of the industry. Reynolds broadened the company's horizons to include specialty cans ranging in size from 5.5 ounces to 32 ounces, offering many growth opportunities.

Reynolds also brought new customers, such as Miller Brewing Company, which Ball did not previously serve. "Overnight, Miller became one of our largest customers," Marty Ruffalo, senior vice president, beverage container sales, points out. Ball worked hard to improve the performance of the acquired Reynolds plants and provide superior products and service to Miller and other customers. Those efforts were recognized in 2001 when Ball received Miller's coveted Supplier of the Year award. Miller gives the award each year to its single best supplier across all products and services. "It was a very proud day for all of us," Ruffalo says. "Even though we had no relationship with Miller until 1998, it took very little time before we became their top supplier of anything."

In every way, the Reynolds acquisition proved a winner for Ball. The Ball publication *Advantage* noted in 1998 that the deal "provided further proof that beverage packaging, even in the mature North American market, can be an attractive business—especially if a company is big, or very good, or both."

Focus on Innovation

Since acquiring the Reynolds plants, Ball has added new production facilities selectively in North America—such as by forming a joint venture with Coors (now Molson Coors) to supply cans and ends to breweries in Colorado and Tennessee and ends to a filling location in Virginia, and by acquiring the Metal Packaging International ends plant in Northglenn, Colorado. Its largest expansion has been in Europe, where can industry sales are still growing, through the acquisition of Schmalbach-Lubeca in 2002.

Low-cost operations and new products are two keys to success in the mature North American market. Ball is well-known for its product quality and efficient operations, but until recently was less well-known for developing new products. "Our customers are beginning to see the can as a brand extension," Michael Herdman, president, metal beverage container operations, emphasizes. "They want their products to stand out on the store shelf and they realize the package is a big part of that differentiation." As a result, Ball has stepped up its research and product development to find new solutions to customers' packaging needs. Previously, research was focused primarily on process improvements in manufacturing. It is now focused on product development as well.

Ball has already introduced several new products, based in part on expanded consumer research, with more in the works. In 2004, it launched a slim take on the traditional 12-ounce aluminum beverage can. "The company hopes the can's sleek look will promote premium product associations and its unique dimensions will make it stand out on the shelf," *Food Engineering* magazine reported.

Also new is Ball's aluminum can for Niebaum-Coppola Winery's sparkling wine Sofia, which is named after Sofia Coppola, daughter of writer-director Francis Ford Coppola, the winery's owner. The 187-milliliter container, nicknamed the Sofia Mini, is the winery's attempt to offer its product in a serving sized right for one person. Further, MSNBC said, "Part of the plan has been to market Sofia not merely as premium wine without a bottle but as a clubgoers' drink: ordered by name, easy to hold at a bar and ideal for resort poolsides where glass is a no-no." The television network quoted Erle Martin, president of the winery, as saying market response has been "overwhelming."

Ball also developed the Straw Can, a cleaner, recyclable alternative to the juice box. It features an easy-opening end enabling the use of a drinking straw. In Europe, Ball is offering "widget technology," which produces a foam head when a beverage is poured. Another European innovation is the use of thermochromic printing inks, which change color in response to changes in the can's temperature—letting the consumer know when the beverage has reached the optimal temperature for drinking.

Innovation in a Mature Market

Ball's objective is to continue to improve the profitability and cash flow of its beverage can business through continuous process improvements, high-quality service and new products that innovatively meet customers' needs.

John Friedery says, "Our challenge is to maximize our performance in a mature market. Nobody makes cans more efficiently than we do, and we intend to get even better. We also have opportunities to develop new products that will help our customers sell beverages. I think we have a very exciting future."

Ball Packaging Europe is a leader in "widget" technology: a foam-generating capsule is activated when the can is opened, creating a foam head that remains stable after the drink is poured. Widget technology was used initially to create the taste and appearance of draft beer from a can and has been extended by Ball Packaging Europe to nonalcoholic beverages such as coffee-flavored drinks.

Can Cup 2004
präsentiert vom Getränkedosenhersteller Ball Packaging Europe
im Rahmen von „Deutschland bewegt sich!"

CHAPTER SIX:

Beverage Cans for a Thirsty World

International expansion is an important element of Ball's strategy. In contrast to the large but mature North American beverage can market, markets in many regions outside North America are less developed and possess good growth potential. Beverage can sales in Europe, for instance, have been increasing at a rate of about 5 percent a year.

Although Ball has been active in international can markets since the mid-1970s, a breakthrough came in 2002 when it acquired Schmalbach-Lubeca, Europe's second largest beverage can maker. The acquisition changed Ball's growth outlook. Prior to the acquisition, Ball derived about one-quarter of its annual revenues from growth businesses, including aerospace and plastic bottles. The company now gets about 40 percent of its revenues from growth opportunities, including the beverage can business in Europe. "We have taken Ball, a good company, and put a little more zip into it," Ray Seabrook says.

All told, international operations generated 23 percent of the company's sales and 34 percent of its earnings in 2004—including the manufacture of beverage cans in Europe, Asia and other regions and, to a much smaller degree, the manufacture of plastic bottles in China.

Early International Growth

Ball began to expand overseas in the mid-1970s by forming can manufacturing joint ventures and licensing its technology. "Our strategy is to develop a leadership position in can making in our domestic market, then export this expertise to emerging world markets," William Lincoln, former executive vice president, metal container operations, explained in 1993.

Ball's first international

Brands such as Heineken beer and Kenco Ice Cappio, a coffee-flavored soft drink produced by Kraft Foods, have been sold in beverage cans supplied by Ball Packaging Europe. The Heineken cans, with embossed shirt buttons, were introduced in Europe to mark the 2003 Rugby World Cup.

Ball began its international expansion in the 1970s and has become one of the world's leading global packagers. Its acquisition of Schmalbach-Lubeca in 2002 substantially increased its presence in Europe. Ball entered China in 1986 and now has seven plants producing metal and plastic packaging there. Ball's joint venture in Brazil, formed in 1995, serves a rapidly growing market.

involvement in beverage cans came in 1974 when it licensed its production technology to Fabricas Monterey of Mexico. By 1997, the company had established joint ventures in Brazil, China, the Philippines, Russia, Taiwan and Thailand, producing nearly six billion beverage cans a year. Its technology was being licensed to manufacturers in Australia, Austria, France, Germany, Israel, Mexico, New Zealand and Sweden, involving another 12 billion cans annually. The combined production of these ventures and licensing arrangements accounted for approximately 8 percent of the beverage can industry's total annual production worldwide, highlighting their important role.

The Chinese market, in particular, caught Ball's eye because of its untapped potential. Today the company has a capacity of 4 billion aluminum cans per year in China in its own plants and in joint-venture plants there, and earns a solid profit, although getting to that position did not come easily. The development of the company's business in China to date divides into three main phases: early promise and rapid growth, disappointment and retrenchment, and the current phase of building a profitable operation from a solid base.

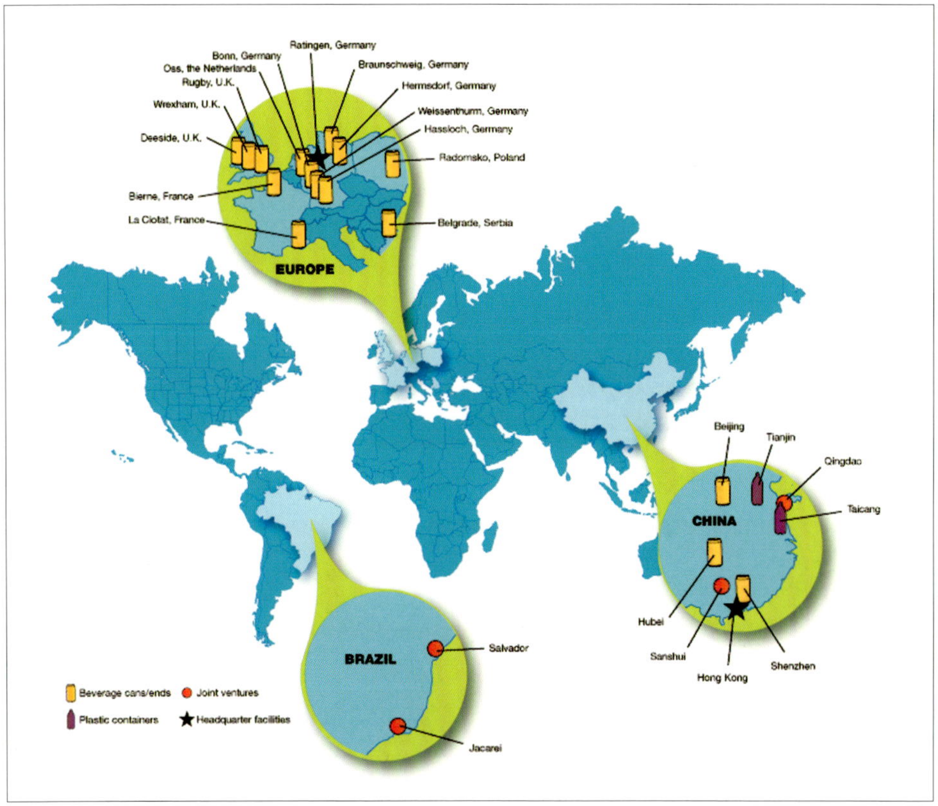

Aluminum beverage cans were virtually unheard of in China until the 1980s. In 1990, *China Business Review* pointed out that beer and soft drinks were sold primarily in glass bottles and that cans were considered "foreign luxury goods" commanding premium prices over other packaging. Looking for new ways to grow, Ball entered the Chinese beverage can market in 1986 by building a joint-venture beverage can factory in Guangzhou, an industrial city with a population of nearly seven million. Its partner was M.C. Packaging (Hong Kong) Ltd. *Chinese Business Review* said the operation was profitable from the start and "received an award as the best new joint venture in China."

Management consultant Tom Peters visited the plant in 1989 and came away impressed. "It has quickly achieved top quality marks from Coca-Cola," he wrote in the *Journal of Organizational Change Management*. "In fact, it was good enough to have garnered the contract to produce all of Coke's cans for the Seoul Olympics in 1988. To top it off, the plant was built from scratch in a startling seven months, half the U.S. average, according to Ball's executives."

Expectations were running high. "The can market in China was growing at a 15 percent compound annual rate, doubling every five years," Dave Hoover recalls. To meet demand, in 1992 Ball entered a second joint venture, FTB Packaging Ltd. It expanded further in 1997 by acquiring majority ownership of M.C. Packaging (Hong Kong) Ltd. Ball and its partners now operated seven plants that produced 50 percent of the aluminum beverage cans in China. Ball also acquired M.C. Packaging's metal food can and plastic bottle businesses.

Expectations were suddenly dashed in 1998 when beverage can demand growth in China came to an

Officials of Ball Asia Pacific Ltd. raise their glasses in toast at an annual dinner. Ball Asia Pacific operates seven plants that make cans and ends for the beverage industry and plastic packaging for the oil industry. In the mid-1990s, then-Ball CEO George Sissel, above, visited some of the company's manufacturing operations in China.

abrupt halt. Not only did exports of cans to other Asian nations fall due to financial market turmoil, but some Chinese soft drink companies started switching from cans to plastic bottles. Faced with industry oversupply, Ball retrenched, closing two plants and taking an after-tax charge of $31 million.

Market conditions only got worse. In 2001, the company closed more plants and wrote down the value of its investment in China by $185 million after tax, the biggest write-down in Ball's history. Today, Ball again has a viable business in China, but on a smaller scale than at its peak. By streamlining operations and finding new customers, the company restored its Chinese operations to profitability in 2002. Ball now makes beverage cans at three plants and two joint-venture facilities in the world's most populous country and plastic bottles at two other plants. The operation, still among China's largest beverage can manufacturers, gives the company a presence that can be expanded if and when opportunities arise. In recent years, growth has returned to the China market, and Ball's operations are once again increasing production.

What lessons were learned from the setbacks in China? Hoover responds, "Dick Ringoen ran this company for a long time, and he used to quote a senator who said there's no education in the second kick of a mule—the point being to learn from your mistakes. We now take a more pragmatic view. We are very tight with

Karlskrone
Premium
PILSENER
F. TSCHOPPE

how we invest our capital because our goal is to generate an adequate return. I'm not going to say we'll never make another mistake, but I don't think we'll make that kind again. We've kissed that pig."

Brazil, the Philippines and Thailand

Ball entered other international markets as well through joint ventures. Aluminum beverage cans were first produced in Brazil in 1989. Ball formed a Brazilian joint venture, Latapack-Ball Embalagens, Ltda., six years later in partnership with the BBM Group and other local shareholders. Each shareholder group owns 50 percent of the venture, which has a can plant in Jacareí, northeast of São Paulo, and an end plant in Salvador, capital of the state of Bahia. The facilities, opened in 1997-1998, were among the most successful startups in Ball's history.

The Brazilian beverage can market has grown rapidly over the past decade and now totals about 10 billion cans a year. Latapack-Ball has a market share of about 16 percent and is one of the top three producers. Customers include Companhia de Bebidas das Américas (AmBev), the leading South American brewer, with such brands as Skol, Antarctica and Brahma; Cervejaria Petrópolis SA, which produces Itaipava, Crystal and Petra beer; and Cervejaria Malta Ltda., which manufactures soft drinks and beer.

Latapack-Ball has delivered strong financial results. "The Brazilian beverage can market is solid and growing, and we continue to generate good results by running efficient plants and serving the needs of our customers," says John Friedery, senior vice president and chief operating officer, North American packaging operations. Bill Gaarder, president of Latapack-Ball from 2000 to 2005 and now vice president, business evaluation, for Ball's North American packaging operations, adds, "We have a highly motivated and talented workforce, one I would stack up against any other workforce anywhere in the world."

Other joint ventures formed by Ball in the 1990s included San Miguel Yamamura Ball in the Philippines in 1994 and Thai Beverage Can in Thailand in 1996. Both continue today as joint ventures.

An Upbeat Stock Market Reaction

Ball's troubles in China arose in late 1997 just as the company was negotiating to purchase Reynolds Metals' can business in the United States. Many Wall Street analysts wondered how Ball, having faltered in China, could manage a large acquisition such as Reynolds.

The Street's attitude changed, however, as the problems in China were resolved, and the Reynolds deal became one of the most successful acquisitions in packaging industry history. In the wake of these successes, when Ball announced on August 30, 2002, that it planned to acquire Schmalbach-Lubeca's beverage can business in Europe, the stock market gave its blessing. Versus a decline of 3 percent when the Reynolds deal was announced, Ball's stock jumped 7 percent when the Schmalbach-Lubeca purchase was disclosed, reflecting Wall Street's new-found confidence in the company.

Frank Tschoppe, opposite, is a production worker at the Ball Packaging Europe plant in Hassloch, Germany. Ball Packaging Europe is the continent's second largest manufacturer of beverage cans, serving many blue-chip customers and brands, as shown below.

European Hotbed of Packaging Innovation

A can is no longer just a can. New ideas from Ball Packaging Europe have made a great package even better. Examples include:

• *Widget technology*: A plastic capsule in the can contains compressed air that is released to provide a lively fizz when the can is opened. Ball has expanded its use from beer to nonalcoholic drinks such as Kenco Ice Cappio, a creamy coffee drink made by Kraft Foods.

• *Aseptic can technology*: Ball Packaging Europe won the 2001 "Can of the Year" international competition for its 0.25-liter aseptically filled can for Modifast, a health drink from Novartis. The beverage is heated briefly to 280 degrees F before being filled in a sterilized can. The process is ideal for milk-based drinks or drinks combining milk and fruit juices, providing longer shelf life.

Ball Packaging Europe pioneered aseptically filled cans. The beverage is heated briefly to 140 degrees centigrade before being filled in a sterilized can. The method makes it possible to package heat-sensitive, noncarbonated beverages in cans and offers a product shelf life of 12 months with consistent quality.

• *FreshCan*: Some drinks contain ingredients, such as vitamins, that degrade in liquid. FreshCan has an internal wedge that houses these ingredients in a dry environment and releases them into the beverage when the can is opened.

• *Active cans*: These cans "communicate" with the consumer, such as by using temperature-sensitive printing inks to display the temperature of the beverage.

• *Photo-quality decoration*: A Ball process introduced in 2004 allows high-resolution photographs and complex color variations to be printed on cans. Customers include Cruzcampo beer in Spain and Jupiler beer in Belgium. Cans with eye-catching graphics stand out on store shelves.

• *Embossed cans*: The first cans with "embossing pure"—which raises the logo or brand name by pressing down the surrounding area—were produced by Ball for Bavaria 8.6 Red beer in 2003.

In 2003, Ball Packaging Europe introduced "embossing pure," an innovative process that accentuates the design details of a can. New printing and embossing techniques permit the creation of unique packaging designs that communicate with the consumer and help sell products. For example, the color inside the mountain icon on the Coors Light cans, opposite, changes from silver to blue to show that the beer is cold enough to drink.

The laser engraving of numbers, drawings or symbols on the underside of tabs is used by beverage companies for competitions and promotions. Coca-Cola, for instance, used printed tabs from Ball Packaging Europe for a 2003 promotional campaign in France in which consumers were able to collect points with the code numbers. The points could be applied toward the purchase of promotional articles.

This time, Wall Street got it right. The Schmalbach-Lubeca acquisition, like the Reynolds acquisition before it, was a big winner for Ball.

Schmalbach-Lubeca

Schmalbach-Lubeca got its start in 1898 when J. A. Schmalbach opened a tinware factory in Brunswick, Germany. Continental Can Company of the United States acquired an interest in the firm in 1935. With support from Continental Can, Schmalbach AG became a packaging innovator—launching the first welded food can made of tin-free steel in 1939, patenting one of the first aerosol cans in 1949 and introducing the first German beer can in 1951.

Further change came in 1967 when Schmalbach merged with Lubeca Werke GmbH to form Schmalbach-Lubeca-Werke AG, one of Europe's largest packaging firms. Continental Can subsequently increased its investment to become majority owner, managing the company as a unit of Continental Can Europe (CCE)—the business Ball sought to buy in 1990.

Following Ball's aborted purchase attempt, Schmalbach-Lubeca endured a series of ownership changes and experienced a period of poor financial results. The firm was acquired in 1991 by German industrial conglomerate Viag AG, with Deutsche Bank holding a minority financial position of 10 percent. Within three years, Schmalbach-Lubeca's earnings fell to break-even, prompting Viag to recruit Hanno Fiedler, who had established an impressive track record in 25 years as an executive with TRW, to improve its performance.

Shortly after Fiedler became CEO in 1996, he and Dave Hoover met at the Drake Hotel in Chicago to get acquainted. Hoover had not given up on Schmalbach-Lubeca and was interested in acquiring the company if it ever came back on the market. "We talked about maybe some day we would have the chance to combine our companies," Hoover says. Over the next seven years, the two met periodically to continue their conversations.

Meanwhile, Fiedler went about the task of improving Schmalbach-Lubeca's performance by selling off money-losing operations and growing its three remaining businesses: beverage cans, plastic packaging, and White Cap bottle caps and closures. By 2000, Schmalbach-Lubeca was again financially healthy and was generating sales of more than $3 billion a year.

> "Why do we have these people here? We don't want to do this. The company is not for sale." With this response, the board essentially endorsed management and confirmed that Ball Corporation would remain independent, as it had been for more than a century.

Ball Explores Potential Mergers

In 2000, Viag merged with another German firm, VEBA, to form today's E.ON AG. The merged company decided to concentrate on its utility business. Consequently, it sold a 51 percent ownership position in Schmalbach-Lubeca to Allianz Capital Partners, a German investment group, and retained 49 percent for itself as an investment.

In the United States, meanwhile, Hoover was involved in a series of actions to chart the future course of Ball. The Reynolds acquisition was proving to be an enormous success, helping drive the company's operating earnings. The stock market, however, continued to ignore Ball's financial performance. In March 2000, as the share prices of dot-com and biotechnology companies soared to record highs in one of the largest speculative binges in stock-market history, the people of Ball could only shake their heads as their own company's share price hovered near its lowest level in two decades.

"The media certainly overlooked us," Scott McCarty, director, corporate relations, says. "One Denver newspaper even took their reporter off of manufacturing to expand the paper's tech-stock coverage. We were a great story but we weren't sexy enough for that time."

Looking back at that strange period in the stock market, Hoover says, "Investors were buying stocks based on promises and hope. They were not buying companies like ours that 'just' made products and earned money."

Ball's inexpensive stock price raised the possibility that another company might bid to acquire it. In January 2001, Hoover succeeded George Sissel as CEO. "Right after I became CEO," Hoover says, "we explored a number of possible mergers and acquisitions. I spent a lot of time on that." None of the proposed deals panned out.

In June 2001, Ball took its large write-off in China. The following month, still exploring options for the future, the board of directors met with investment bankers to consider a merger. The board concluded that such a transaction essentially would sell the company. According to Hoover, "The board finally said, 'Why do we have these people here? We don't want to do this. The company is not for sale.'" With this response, the board essentially endorsed management and confirmed that Ball Corporation would remain independent, as it had been for more than a century. Hoover says, "I had spent my first six months as CEO negotiating the possible merger of the company and writing down the value of our business in China. The board's endorsement was a huge vote of confidence and very motivating to our management team."

Ball Packaging Europe has plants in Germany, France, the Netherlands, Poland, the United Kingdom and Serbia. It has an extensive presence both in the traditional core markets of western Europe and the significant growth regions of eastern and southern Europe.

Schmalbach-Lubeca Comes Back on the Market

The company now set its sights on acquiring Schmalbach-Lubeca. Allianz, the new majority owner, was already looking to sell its investment and take a

Younger consumers in Europe like the on-the-go advantages of beverages in cans.

profit. Hanno Fiedler was instructed to find the best solution for the owners, the business and its employees.

Fiedler says his first choice was to try to have the company sold to Ball, because he knew Hoover personally and believed that Ball would value Schmalbach-Lubeca and invest in the continued growth of its business. He did not want to sell to another equity group, which would simply own the company until it could dispose of it at a profit.

But there was a sticking point: Ball only wanted to buy Schmalbach-Lubeca's beverage can business, not its plastic container and closure businesses. Although Fiedler preferred to sell the company to Ball, equally important was his desire to sell the entire firm to one buyer. He could not find a single buyer, however, so he persuaded the German owners to sell part of the company—its plastic container and closure businesses—to Australian packaging firm Amcor Limited. That opened the door for Ball to acquire the remainder of Schmalbach-Lubeca: the beverage can business it had wanted for more than a decade, ever since it failed to acquire it when it had been called Continental Can Europe.

"The Schmalbach acquisition suddenly became much more doable for Ball, much more interesting," Fiedler says. The issue then became price. "It was my task," Fiedler says, "to convince Ball to pay a realistic and fair price and to convince the owners not to exaggerate their price expectations. So it was an interesting period in my life."

Ball sent teams of people to Europe to examine Schmalbach-Lubeca's operations and determine their value. After completing this due diligence, and following several months of negotiations, Ball and the sellers—Allianz and E.on—reached a deal on August 29, 2002. The agreed-on price: 922 million euros, equivalent to about $948 million. As the buyer, Hoover describes the price as "attractive," adding, "We didn't pay too much." His friend and colleague Fiedler, speaking for the seller, says the price was "fair," and adds that the sellers were satisfied that they received a good price and the business was assured of having an excellent new owner focused on packaging. The deal closed on December 19, 2002, whereupon Schmalbach-Lubeca became part of Ball.

Fiedler had wanted to ensure a stable future for Schmalbach-Lubeca and its people—which is why he preferred the sale to Ball. "I was able to lead our people to a safe harbor," he says. "We were now part of a true packaging company. People do change, but there is no reason why Ball would spin off such a European activity unless they were to give up can making. And when you look at Ball's history, there is no reason they should get out of can making."

"A Great Strategic Fit"

In August 2002, Hoover told investors in a Web cast, "This acquisition is a great strategic fit for us, giving Ball a strong presence in Europe. The European beverage can market has experienced growth and is the second largest market in the world for beverage cans."

Based in Ratingen, Germany, Schmalbach-Lubeca was one of Europe's premier makers of aluminum and steel cans. (Unlike North America, where steel beverage cans are no longer produced, they still have an active market in Europe.) Its 12 plants throughout Europe produced 12 billion cans and ends a year, representing about a one-third share of the European beverage can market.

The Reynolds acquisition had been attractive to Ball because of the opportunities it afforded for cost-cutting synergies. Schmalbach-Lubeca was attractive, by contrast, because of its growth opportunities. Schmalbach was already highly profitable and well run, so there was no need to quickly close plants or reconfigure operations. It contributed to Ball's profits from the day it was acquired.

One of the few negatives to the acquisition was the impact of Germany's new 25-eurocent deposit imposed on many one-way containers, including beverage cans. Because Germany lacked a nationwide system for returning containers, consumer demand for beverage cans there declined sharply soon after Ball acquired Schmalbach-Lubeca. Nonetheless, the law was already in the works when Ball was negotiating to buy Schmalbach-Lubeca, so the projected impact was well understood and was factored into the price Ball paid.

Fiedler joined Ball as executive vice president and a member of its board of directors. He continued as chairman and chief executive officer of the former Schmalbach-Lubeca operations, renamed Ball Packaging Europe. Jan Driessens continued as the unit's president.

Fiedler notes that American acquirers of non-U.S. businesses often replace the acquired company's entire senior management team to gain better control. "A company might send 25 people from its headquarters to staff certain key jobs in an acquired company to find out what's going on," he says. Ball, by contrast, sent three people—a new chief financial officer, a new treasurer and a technical center vice president. Fiedler and Driessens have continued to lead Ball Packaging Europe since the acquisition by Ball, providing continuity and expanding the business both geographically and through the introduction of new products. In 2005, Fiedler announced his intention to retire at the end of the year, when he will become chairman of the supervisory board for Ball Packaging Europe. He will continue as a member of the Ball Corporation board of directors. Driessens plans to retire in mid-2006 and will continue as a consultant to Ball Packaging Europe.

Since the acquisition, Ball Packaging Europe has continued its profitable growth. Moreover, looking to the future, a plant built in Serbia in 2005 is positioned to serve the increasing demand for beverage cans in southern and eastern Europe. Driessens points out, "With Serbia as its center, we will have a regional market of 60 million to 70 million people." He expects double-digit growth in beverage can demand in central and eastern Europe for the next several years. Noting the sizable potential of the market, he adds, "In western Europe, an average of 70 beverage cans per capita are consumed annually. The figure for eastern Europe is as low as 13." He also sees possibilities for new plants to serve growth regions not only in Europe but eventually in Asia and Africa.

With its operations in China and other nations, and most particularly with its acquisition of Schmalbach-Lubeca, Ball has emerged as one of the leading beverage can manufacturers in the world, not just in North America. Ball in the United States and Ball in Europe now "steal shamelessly" from each other, in Fiedler's and Driessen's words, borrowing and implementing the best ideas. Ball Europe, for instance, is a leader in developing new packages, some of which have been adopted for the U.S. market. Ball in the United States, conversely, is well-known for its manufacturing technologies, some of which have been exported to Europe.

Hanno Fiedler, at right in photo above, chairman and chief executive officer of Ball Packaging Europe, speaks in 2004 at the groundbreaking for a beverage can plant in Belgrade, Serbia. The facility makes beverage cans for the Serbian market as well as for export to Romania, Bulgaria, Hungary, Slovenia, Croatia and other nations. Second and third from left, respectively, are Serbian Prime Minister Vojislav Kostunica and Deputy Prime Minister Miroljub Labus.

CHAPTER SEVEN:

Metal Food Cans: Keeping Foods Fresh, Safe and Convenient

Metal food cans have a long history of providing convenience and other benefits to consumers. They were first manufactured in England in 1813 and came into widespread use during the U.S. Civil War as a way to furnish meals to soldiers in the field. By the early 20th century, canned food was a staple in the American kitchen.

Brian Cardno, Ball's president, metal food container operations, calls the metal food can "the perfect container." He says it offers many advantages over other packaging, including generally lower costs, convenience and long shelf life as well as preservation of food flavor and nutritional content. What's more, food cans "are recyclable, don't break and are probably the safest food package available," he says.

Canned food is not limited to everyday meals in the home. "Even top chefs like to open a can," the *Denver Post* proclaimed in a May 2004 article that praised the versatility and wholesomeness of canned food. The newspaper quoted several professional chefs about their use of canned food in recipes.

The principle of the food can is simple but effective. Food is preserved by applying the same methods homemakers have used for years to preserve fruits and vegetables in glass canning jars: the food is heated to kill bacteria and germs and is then sealed in a container. Studies at the University of Illinois and University of Massachusetts have shown that canned foods are similar in nutritional value to their fresh-cooked

Metal food cans are cost-effective, convenient and recyclable. They protect the taste and nutritional content of fruits, vegetables, fish, soups and other foods in tamper-evident containers. Ball manufactures food cans at 11 locations across North America and supplies cans for products such as Allens Cut Italian Green Beans, Bush's Beans, Chef Boyardee Spaghetti & Meatballs, Red Gold Premium Tomatoes and Ship Ahoy Salmon.

In 2003, Ball installed the world's fastest food can line at its Milwaukee beverage can plant. The Milwaukee facility is one of three within Ball where food and beverage cans are produced in the same location. The food can line is able to produce two-piece, 14-ounce recyclable cans at the rate of 4,000 per minute, or about 1.2 billion every year.

and frozen counterparts. Moreover, this nutrition is retained for at least two years from the date of processing. At temperatures of 75 degrees Fahrenheit and below, the shelf life of canned food is almost indefinite. In a famous example, a can of roast veal and gravy from Edward Parry's third voyage in search of a Northwest Passage in 1824 was opened and examined more than a century later in 1939. Laboratory tests revealed that the contents retained most of their nutrition and were still edible.

Growing in a Mature Industry

Ball manufactured its first steel food can in 1988 and is today the second largest player in the large but mature North American food can market, supplying cans to leading food companies such as Allen Canning, Bush Brothers, ConAgra, Dean Foods, Red Gold, Ross Labs and Trident Seafoods, among many others.

Although industry demand has remained essentially unchanged for more than a decade at about 31 billion cans a year, reflecting inroads from other packaging materials such as plastics, Ball has grown by aligning with customers that are expanding, investing in new production facilities and acquiring other can manufacturers. In 2004, Ball produced 6.5 billion food cans, up from 4.6 billion eight years earlier. Food cans accounted for about one-seventh of the company's total revenues in 2004.

Ball makes two-piece as well as three-piece steel food cans. The three-piece can has a separate top, bottom and side wall, whereas the bottom dome and side wall of a two-piece can are made from a single piece of metal. Because the metal in a two-piece can is drawn to a thinner gauge than in a three-piece can, two-piece cans typically have a low height-to-diameter ratio. For strength, taller cans are normally of three-piece construction.

Because of the margins in the food can business, low-cost operations are essential to success. Cardno says Ball's food can plants across North America rank among the most modern and efficient in the industry. The company continues to invest in production improve-

Ball makes three-piece welded steel food cans, pictured here, as well as two-piece cans, meeting customers' preferences and needs for either type.

ments, such as a high-speed food can production line installed in 2003 at its beverage can factory in Milwaukee. Costs were reduced by building the line in an existing plant. Moreover, John Friedery says the facility is "certainly the most modern two-piece steel food can line in the world." The line enabled Ball to serve a customer that reduced its self-manufacture of cans. As a result, the new facility—with an annual capacity of 1.2 billion recyclable steel cans—did not add net capacity to a mature industry.

By combining Heekin with Ball Packaging Products Canada, Ball became one of the largest suppliers to the combined U.S.-Canadian food can market and strengthened its market position in the Midwest.

Entering the Business in 1988

As with many of the moves that Ball has made successfully, the company entered the food can business by quickly seizing on an unforeseen market opportunity—in this case, two of them. In 1983, Canadian entrepreneur Gerald Schwartz founded Onex Corporation to acquire companies in leveraged buyouts. In one of his earliest deals, he bought American Can Company's Canadian packaging business, took it public and renamed it Onex Packaging Inc. The operation manufactured steel food cans, aluminum beverage cans and household and industrial cans for the Canadian market.

By 1987, however, Onex Packaging was struggling, and Schwartz decided to find a strong American partner to help reduce costs and improve productivity. He turned to Ball, which made aluminum beverage cans and was looking for opportunities in the food can industry. In 1988, Ball and Onex Corporation created a joint venture, Ball Packaging Products Canada, Inc., which bought Onex Packaging and took it private in a leveraged transaction. Ball and Onex Corporation each owned 50 percent of the new firm.

Ball proceeded to take "a lot of costs out of the business," *Investor's Business Daily* reported. But food can prices in Canada began to fall when the 1988 Canada-United States Free Trade Agreement opened the doors to duty-free U.S. imports. On top of that, interest rates rose sharply, adding to the venture's interest expense. Beset by this deadly combination of lower prices and higher interest costs, the venture was unable to pay its lenders and was placed in receivership in March 1991. After four weeks of legal proceedings, Ball paid the banks a discounted amount for the debt and owned the company outright. Others had the right to bid, but Ball offered the most: $111 million for debt with a face value of $171 million.

Ball worked relentlessly in the ensuing years to reduce costs so that the venture's plants would become fully competitive with can factories in the United States. As it achieved its cost goals, it began to expand sales across the border to U.S. food packers.

Ball Packaging Products Canada, the number-one supplier of food cans in Canada, became the base upon which Ball built its food can business in North America.

Acquiring Heekin Can

In 1990, *Fortune* magazine wrote about an Ohio-based food can manufacturer, Heekin Can, Inc. "The company was founded in 1901 by a tea and spice merchant who was appalled at the prices he was forced to pay for cans," the magazine said. "He started making his own and soon canned the less profitable teas and spices."

The magazine also wrote, "Who needs glamour in these tough economic times? Not this Cincinnati can company—and not the buyers of Stokely green beans, Freshlike corn, Van Kamp baked beans, and other products sold in Heekin containers. When budgets are tight, consumers can save a buck by eating at home and choosing less expensive canned veggies rather than fresh or frozen ones. Heekin's chief financial officer, Perry Schwartz, says canned food sales usually rise in a recession, making his industry 'about as bullet-proof as you can get.'"

In 1993, Ball—seeking to continue its growth in metal packaging—acquired Heekin and its 11 manufacturing plants for $213 million, including the assumption of debt. The acquisition more than doubled the size of Ball's food can business. "As one of the largest food can manufacturers in the Midwest," Ball CEO Del Davis wrote, "Heekin presented one of the few opportunities available to us to become a significant player in the metal food can market with a single acquisition."

By combining Heekin with Ball Packaging Products Canada, Ball became one of the largest suppliers to the combined U.S.-Canadian food can market and strengthened its market position in the Midwest.

Since then, Ball has continued to develop its food can business. In 2000, it formed Ball Western Can Company, a joint venture with food giant ConAgra. The venture assumed operation of ConAgra's food can plant in Oakdale, California, with opportunities to supply not only ConAgra but other West Coast food companies as well. "This agreement gives Ball a food can presence in an important market and it does so without investing much capital," John Hayes, then vice president, corporate strategy and development, said. Four years later, Ball acquired full ownership of the plant with a long-term agreement to continue to supply ConAgra's can needs on the West Coast.

With this transaction, Ball cemented its place as the second largest manufacturer of metal food cans in North America, a position it continues to hold as of 2005. Silgan is the largest food can manufacturer, followed by Ball and then Crown Holdings, but Ball has gained share in a consolidating industry.

In 1993, Ball acquired food-can manufacturer Heekin Can, Inc., founded in 1901. Clif Reichard, Ball's former customer relations director, packaging operations, donned this unusual outfit, below left, in 1994 for a Halloween party at food-can customer Morgan Foods in Austin, Indiana.

Ball
Packaging Products Canada, Inc.

Lower Costs, New Products

What's next for Ball in food cans? "Food cans are a mature business," Cardno emphasizes. "But one of the ways to make a nice living in a mature business is to be the low-cost producer." He adds, "We continue to amaze ourselves in finding ways to get costs out of the product. And that has been the sustenance of the business, really."

Costs have been reduced in many ways, such as by speeding production lines, reducing the amount of metal in cans, operating plants around the clock to maximize return on investment and consolidating facilities by closing less productive plants. However, as manufacturing costs have gone down, so have food can prices. "Can prices over the past decade have increased at a rate well below inflation and in some cases have declined in absolute terms," Cardno says. Consequently, productivity improvements and cost efficiencies remain a priority.

Cardno also sees attractive opportunities for new products, such as resealable coffee cans and microwaveable metal containers that allow canned foods to be heated in their packages says Hayes, former vice president of corporate strategy, marketing and development who in 2005 was named executive vice president of Ball Packaging Europe. "Food cans are a commodity, but they are also about convenience. We have increased our research to develop new convenience products that will support our food can business and meet the needs of our customers." He says Ball is seeking in particular to develop innovative can features that will help food companies differentiate their products via packaging. "Our goal is to meet the needs and desires of consumers, and do so with features that are cost effective for our food-company customers," he says.

The metal food can business is tough and competitive, yet Ball continues to succeed—serving its customers with quality products while generating strong cash flow with some opportunities for growth.

Ball has a strong position in salmon packaging, supplying nearly all of Canada's and much of Alaska's market. Willie Michaluk, opposite, was press mechanic for Ball Packaging Products Canada's metal food container plant in Richmond, B.C., in the early 1990s. Above, a variety of vegetables come packed in Ball's food cans in North America.

BALL
M WINE BOTTLE
87 ML BOTTLES
BALL
PREMIUM
WINE BOTTLE
187 ML

CHAPTER EIGHT:

Versatility in a Bottle: PET Plastics

Ball has repeatedly shown its ability to offer new packaging materials and change with the needs of its customers. The switch from glass to plastic is a prime example. In November 1994, months before it sold majority ownership of its glass business, the company announced that it was entering the PET (polyethylene terephthalate) plastic bottle business. PET is a clear, versatile plastic used for the manufacture of soft drink and water bottles and other products.

Demand for plastic bottles was growing dramatically. In its 1994 annual report, the company said the total U.S. plastic container market was expected to reach $5.5 billion in 1994, surpassing sales of glass containers for the first time, and was projected to advance to $10 billion by 2003. "We realize," the report noted, "that PET plastic containers are increasingly part of the packaging mix."

Ball entered the plastic container business in part at the urging of longtime customers. "They told Ball, 'We think you're a great company, and if you were in PET we would do even more business with you,'" Mike Vaughn, vice president, packaging innovation, says.

By coincidence, U.S. production of aluminum beverage cans peaked at 103 billion in 1994, the year in which Ball announced its diversification into PET. Since then, beverage can demand has declined slightly, although Ball—with its strong ongoing commitment to the beverage can market—has done well by meeting the needs of its beverage can customers, continuously reducing costs and leading the industry's consolidation. Beverage cans continue to have a bright future even as plastic has captured

Ball, a leader in the PET plastic container business, makes a broad range of products, including water bottles and hot-fill juice bottles. To address growing needs for use in stadiums, nightclubs and outdoors, Ball also makes plastic bottles for beer

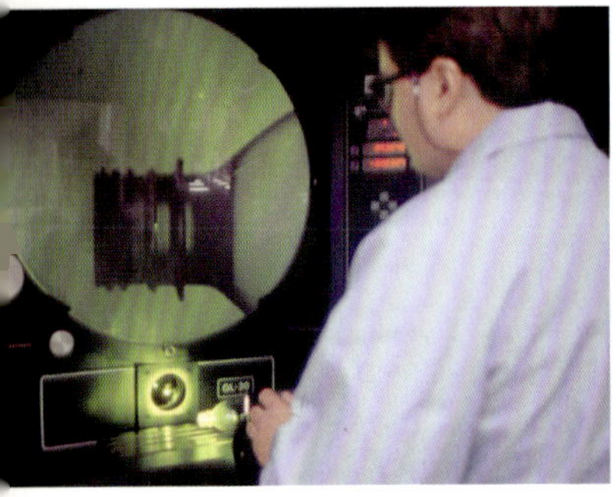

To measure the attributes of plastic bottles, technicians in Ball's PET operations use a comparator, which projects an image on to a screen with a numerically scaled grid. Above, Norm Van Horn, director, quality systems and manufacturing support, uses the device to check the dimensions of closure threads. Ball manufactures PET plastic bottles in a two-step process. First comes production of preforms, opposite, which look like test tubes. The preforms are then fed into a machine that heats, stretches and blows them into the shape of bottles, near right.

much of the growth in packaging, just as Ball anticipated.

Ball has steadily built its position in the PET container market through the startup of large, modern plants and strategic acquisitions and is now able to meet customers' packaging preferences, whether for metal or plastic. As of 2004, the company manufactured about 5.5 billion plastic bottles a year and was the sixth largest U.S. producer in a fragmented market with many participants. The PET container business accounted for about one-twelfth of the company's worldwide sales in 2004.

Invention of the PET Beverage Bottle

The PET beverage bottle was invented by Nathaniel Wyeth, who was born into one of America's most famous families of artists—including his father N. C., brother Andrew and sisters Henriette and Carolyn Wyeth.

Nathaniel Wyeth, who spent most of his career at DuPont Company, was an artist of a different sort—in engineering. One day in 1967, he wondered out loud why plastic was not used for carbonated beverage bottles. A colleague replied that the bottles would explode. Questioning the idea, Wyeth bought a plastic detergent bottle. Filling it with ginger ale, he found that the bottle stretched but did not rupture. He then spent nearly a decade persistently experimenting with various plastics and production methods. In time, he developed a pre-form mold, resembling a large test tube with a crisscross pattern inside it, that aligned the plastic's molecules to reduce stretching. The resulting amorphous plastic pre-forms could then be blow-molded to form bottles. Wyeth's two-step production process is still the industry norm. After testing thousands of resins, he found that PET provided the best results, offering maximum strength with minimum stretching.

Among other advantages, PET bottles are lightweight and recyclable, can be molded into distinctive shapes and do not break when dropped. Pepsi introduced the first soda in PET bottles in 1977 in western Michigan, followed soon by Coke. "It has been a wild ride ever since," *Food Technology* magazine observes.

Early Growth

Ball moved quickly to enter the PET business following its 1994 announcement. It hired an experienced management team—led by Larry Green, who joined the company from bottle maker Constar International—and built a pilot line and research and development center in Atlanta. It soon constructed its first manufacturing facility in Chino, California, followed by plants in Baldwinsville, New York; Ames, Iowa; and Delran, New Jersey.

"We started with four people and five pencils," Vaughn jokes. "We built four plants in four years, spent $300 million and hired 600 people. We went from zero sales to about $250 million in that same time frame."

Ball chose primarily to build new plants rather than acquire existing facilities because PET production machinery had recently and significantly improved. "We feel that our

Ball manufactures PET plastic bottles at large, modern plants in Ames, Iowa; Baldwinsville, New York; Chino, California; Delran, New Jersey; and Watertown, Wisconsin. Expansion of the Ames plant, left, in 2002 added two new production lines and 400,000 square feet of warehouse space, making it Ball's largest plastic bottle facility. Jeff Thede, assistant production manager, and Tim Malott, production manager, inspected bottles in 2003.

Larry Green has led Ball's plastic container operations since their inception in 1994. "Plastic bottles are a young business with a great deal of growth potential," he says. Pepsi-Cola is the largest customer of the company's plastic bottle business.

modern technology allows us to be a low-cost, high-quality producer," Green told *Packaging World Magazine* in 1997.

Success, however, did not come easily. The Chino plant produced its first bottles in October 1995, but then the problems started. It took longer than anticipated for the facility to operate efficiently at full capacity. Numerous technical problems plagued the operation. While it is not unusual for new factories to be slow to reach full capacity, Chino went beyond the norm. Green and other senior executives spent many hours there working with engineers and other plant personnel to get the new facility running smoothly. One challenge was to learn how to put labels on plastic bottles, something Ball people had never done before. "It was a struggle," Green says, "but we worked hard and got our costs down and our quality up." The story eventually had a happy ending. In May 1997, *Plastic News* reported that, after months of tribulation, Chino had become a "showcase," adding, "There's no doubt that Ball Corp.'s move into PET container manufacturing is proceeding at a rapid clip."

Meanwhile, the Ames and Baldwinsville facilities began operations in 1996, Delran in 1997.

Green has led the business since its inception. He says an initial objective was to sell PET containers to Ball's glass container customers, but that strategy changed

Pushing the Limits of Plastics

PET plastic is one of the major growth stories in today's world of consumer packaging. Ball CEO Dave Hoover says, "Every time I go to the grocery store there is something else sitting there in a plastic bottle that used to be in glass."

PET has many advantages. It can be readily shaped and colored to create unique packaging; it is lightweight, low-cost, resealable, shatter-resistant and taste-neutral. Furthermore, specialty coatings and barriers (such as nylon or gas) can be embedded in or between layers of plastic to meet particular needs, such as blocking light and oxygen from entering the container and degrading the contents.

Ball has staked its growth in PET bottles on proprietary products that take advantage of plastic's versatility. The *Daily Camera*, a newspaper in Boulder, Colorado, noted in 2004 that Ball is

PET offers great design freedom. It can be shaped into many sizes, shapes, colors and designs.

"pushing the limits of … cans and plastics—while making sure they're cheap enough to make on a wide scale—to find better, flashier, easier-to-use containers."

Ball makes many different sizes and types of plastic bottles for alcoholic beverages. They include 12-ounce beer bottles, such as those produced for Anchor Brewing's Anchor Steam brand, and 187-milliliter wine bottles, such as those sold in four-packs by Sutter Home Winery.

Ball also makes PET bottles for flavored liquors, vitamin-enhanced water and other beverages and has supplied environmentally friendly plastic beer bottles to Molson Coors and Anchor Brewing of San Francisco.

In September 2004, Ball introduced its Heat-Tek® line of PET bottles that can be filled with liquids heated to 185 degrees Fahrenheit, which is required for juices and other specialty drinks. The company is focusing on Heat-Tek and multilayer packages to grow its custom PET business.

Ball manufactures single-serve and multi-serve plastic bottles, opposite, with grip features and other standard and custom designs. Whether on a plastic bottle or other product, the Ball signature, below, remains as much a signature of excellence today as it was a century ago.

when Ball divested its glass container business. As a result, the plastic container team began marketing to the company's beverage can customers as well as to other companies that were interested in plastic packaging.

Support from management, including retired CEOs Ed Ball and John Fisher, was critical to success. Speaking of Ed Ball (who was approaching 90 when the company entered the PET business) and Fisher (who was approaching 80), Green says, "Whenever we said we needed them to see a customer with us, they would do so without question."

Ball Corporation at that time maintained a guest house in the former mansion of co-founder William Ball in Muncie. Customers and prospects visiting Muncie were invited to stay there. "I'll never forget the time Ed and John couldn't make it to the guest house to have dinner with a customer because they were attending a charitable function that evening in Indianapolis," Green says. "But they agreed to have breakfast with the customer the next morning. And they showed up at 6:45 a.m. even though they had not returned home from Indianapolis until well after midnight."

He says "customers loved meeting Ed and John," who would talk about the history of the company, its commitment to customers and its vision for the future. Chatting with Ball and Fisher, as well as with members of the company's active senior management team, helped convince customers and prospects that Ball Corporation—a newcomer to the PET bottle business—would be a reliable supplier. In addition, Green says, "Beverage companies knew Ball was a quality manufacturer of cans and they counted on us to do the same in PET."

An early breakthrough occurred when Ball negotiated a long-term contract to supply PET bottles to PepsiCo. That contract, announced in April 1995, made headlines in the business press. *American Metal Market* even stated that "two shots were fired across the aluminum industry's bow this week—one for effect and the other a warning, and both bracing the industry's biggest market: the beverage can." One of these "shots" was the Ball-Pepsi announcement, the other an announcement by aluminum manufacturer Pechiney that it would no longer invest in the growth of a subsidiary's beverage can business because of the competitive threat from PET containers.

The threat to aluminum cans was less severe than it may have seemed at the time. Nonetheless, PET's share of the packaging market has grown.

Pepsi-Cola continues today as the largest customer for Ball's PET containers. Ball supplies various Pepsi locations not only with PET bottles but also with aluminum beverage cans and is one of that company's leading outside suppliers of any product or service.

In addition to building new plants, Ball came up with the idea of buying the bottle-making capacity of regional soft drink companies and arranging long-term contracts to supply the companies' bottle needs, in this way adding important new regional customers. In 1997, for example,

San Francisco's Anchor Brewing Company, the maker of Anchor Steam beer, is among Ball's customers for plastic beer bottles.

it purchased two bottle-making plants from Honickman Group, a billion-dollar private bottler and distributor that has Pepsi-Cola and Canada Dry franchises as well as private-label facilities in the Northeast. Ball closed the plants, which were relatively small, and built a larger, more efficient plant of its own in Delran, New Jersey, to help supply PET bottles to Honickman.

In 2001, the company made a similar arrangement with Pepsi bottler Wis-Pak, Inc., in the Midwest. It acquired Wis-Pak's bottle-making facilities in Sioux City, Iowa, and Watertown, Wisconsin. The Sioux City plant was closed while the Watertown plant was retained, adding to Ball's strong market presence in the Midwest. Today, Ball supplies 100 percent of Wis-Pak's PET container requirements from Watertown and Ames.

Focus on New Products

When Ball entered the PET business, demand was growing dramatically. Since then, while demand for PET containers has continued to increase, pricing has weakened due to industry overcapacity.

In 2004, Ball announced it would not make any major new investments in the commodity side of its plastics business until the market situation improved. "There are too few customers for these containers and too many competitors selling for too low a price," Dave Hoover said. He believes "restructuring or consolidation" needs to take place in the industry to enable bottle manufacturers to earn satisfactory returns on investment.

Ball continues to invest in manufacturing process improvements and new products. Until recently, Ball concentrated mainly on the manufacture of commodity-type PET packaging, including soft drink and water bottles. It is now seeking to expand its sales of proprietary plastic packaging products. "Currently we have 66 active projects in the new product area," Vaughn said in 2005.

Plastic packaging continues to capture market share from glass. One of the biggest opportunities is to develop plastic packaging for aseptic foods and bever-

"We didn't do it ourselves. It was the relationships this company had with its customers that allowed us to start a business and have enabled it to grow."

Increased demand for plastic containers is being driven primarily by conversion from glass to PET. Ball is targeting its growth to propriety products such as plastic beer bottles, left, single-serve juice containers and bottles for hot-filled beverages.

ages. "Aseptic foods and beverages have been a U.S. trend waiting to happen for more decades than most Americans have been alive," *Food Engineering* magazine wrote in 2004. When consumers drink a bottled beverage, they expect it to taste fresh and be the same no matter where they buy it. However, meeting consumer expectations has traditionally been something of a challenge when it comes to juice-based products, which are made from highly perishable and sometimes inconsistent ingredients. To address beverage companies' needs, Ball has expanded its line of Heat-Tek® PET plastic bottles for hot-filled beverages, including juices, teas, sports drinks, vitamin-enhanced water and other products that are filled at temperatures up to 185 degrees Fahrenheit to ensure their safety, maintain their taste and extend their shelf life. "Consumer demand for healthier beverages and conversion from glass to PET continue to drive market growth," Allan Benson, Ball's director, custom PET, says.

Other opportunities include plastic bottles for flavored alcoholic beverages and beer. Beer is already being sold in plastic bottles at venues such as Miami's Pro Players Stadium, Bank One Ballpark in Phoenix and Arco Arena in Sacramento. In Las Vegas, an ordinance was proposed to ban glass bottles on New Year's Eve and other special times. The benefits of plastic include safety, convenience and reclosability. Consumer acceptance is growing.

Ball's plastic beer bottles look just like their glass counterparts. However, it has proprietary "barriers" between layers of plastic to keep carbonation in, keep oxygen out and maintain flavor. Ball is actively marketing the product.

Ball has come a long way in the PET plastic container business since starting from scratch in 1994. Green says, "People ask me all the time, 'I bet it was fun starting the business. Just three of you and not having a single customer. Now your sales are $400 million. You guys really did a great job.' But I say, 'Time out. We came to Ball and did that. We didn't do it ourselves. It was the relationships this company had with its customers that allowed us to start a business and have enabled it to grow.'"

CHAPTER NINE:

Aerospace: Reaching for the Stars

On May 10, 1957, a one-paragraph story in the *New York Times* reported that the Ball Brothers Company, "manufacturer of glass, metals, rubber, plastics and electronics products, has formed a new research subsidiary in Boulder. The subsidiary, which will operate under the name of Ball Brothers Research Corporation, will conduct fundamental research and applied research in electronics, astrophysics, optics, servo mechanisms and other instrumentation."

From that modest beginning, Ball Aerospace & Technologies Corp., as it is known today, has evolved into a highly successful aerospace business. "Our core strength is solving very complicated technical problems with very talented and dedicated people," Ball Aerospace President and CEO Dave Taylor asserts. Ball Aerospace has been engaged to provide solutions as important and demanding as fixing the blurred vision of the Hubble Space Telescope and providing secure voice, fax and data transmission systems for Air Force One. The firm prides itself on a culture that puts a premium on science and innovation. More than one-quarter of its approximately 3,400 employees have master's degrees or doctorates.

Ball Aerospace, Ball's oldest business now that the corporation no longer makes glass, is a successful niche player in the huge field of space exploration. Smaller than its giant rivals, it is best known for space-qualified sensors and instruments, currently in strong demand. "The market is pointing right at us," Taylor says. Clients include NASA, the U.S. Department of Defense and other government and private-sector organizations.

Ball Aerospace has grown rapidly in recent years, nearly doubling its sales and earnings from 2000 to 2004. Sales were a record $653 million in 2004 and the year-

Ball Aerospace is helping answer some of the universe's most vexing questions. Deep Impact, opposite, placed a spacecraft in the path of a comet, with the resulting collision ejecting material that may provide clues about the solar system's origin. Ball developed several instruments for the Spitzer Space Telescope, which can peer into clouds of dust and gas such as the nebula, left, that surrounds Cassiopeia A, a once-massive star that died in a supernova explosion 300-plus years ago.

Ball Aerospace built three scientific instruments and the communications equipment for the CALIPSO satellite, to study the Earth's radiation balance. Many climate models predict significant global warming, but confidence in these predictions is low. CALIPSO will measure the impact of aerosols and clouds to help policy makers make informed decisions about global climate change.

end backlog reached a new high of $694 million. The workforce grew by a net 380 employees in 2004 alone. While accounting for about 12 percent of Ball Corporation's revenues in 2004, Ball Aerospace more than pulls its weight in return on investment and contribution to earnings. As Ball Chairman, President and CEO Dave Hoover says, "Aerospace makes money, we're very good at it and it's growing faster than most of our other businesses."

The Space Race Leads to Early Growth

Ball Aerospace was founded a year before NASA was established and it has grown side by side with the U.S. space agency for nearly half a century. The space race got off to a sudden start in October 1957 when the Russians launched Sputnik, the world's first manmade satellite. In 1958, Congress created NASA to lead America's space effort. Ball Aerospace had several of the technical capabilities NASA was looking for, and the two have been working together ever since.

"We were a company that was started to explore space," Donovan Hicks, who led Ball Aerospace from 1988 to 1996, said in the mid-1990s. By contrast, he said, "Most of the large aerospace companies that found their way into this business were already big players in the defense community and gravitated to space as an offshoot of their other business."

Ball Corporation was struggling in the glass business in the 1950s, so the emerging field of space exploration presented an interesting opportunity to diversify into an industry with dynamic growth potential. The *Denver Post* notes, "There was no need for Ball to own an aerospace company, but the new division had another value that persists to this day. Its scientists were really smart, and they could earn reliable income by designing space gadgets for the government."

At its beginning, Ball Aerospace was headquartered in Boulder, Colorado, and became a magnet for scientists and engineers from the nearby University of Colorado (CU). Early employees with connections to CU, such as Ruel C. "Merc" Mercure Jr. and O. E. "Pete" Bartoe Jr., were attracted by the prospect of solving complex new technical problems. Mercure, who has a doctorate in physics from CU, spent 24 years at Ball in a variety of positions, including president of Ball Brothers Research Corporation, Western group vice president and group vice president-technical products. He later served on Ball's board of directors.

Bartoe earned bachelor's and master's degrees in mechanical engineering from CU and has had a lifelong interest in aviation. He joined Ball Brothers Research Corporation at its founding and led the firm from 1970 to 1973.

Many of the first employees were offbeat, creative people who enjoyed the firm's relaxed but rigorous work

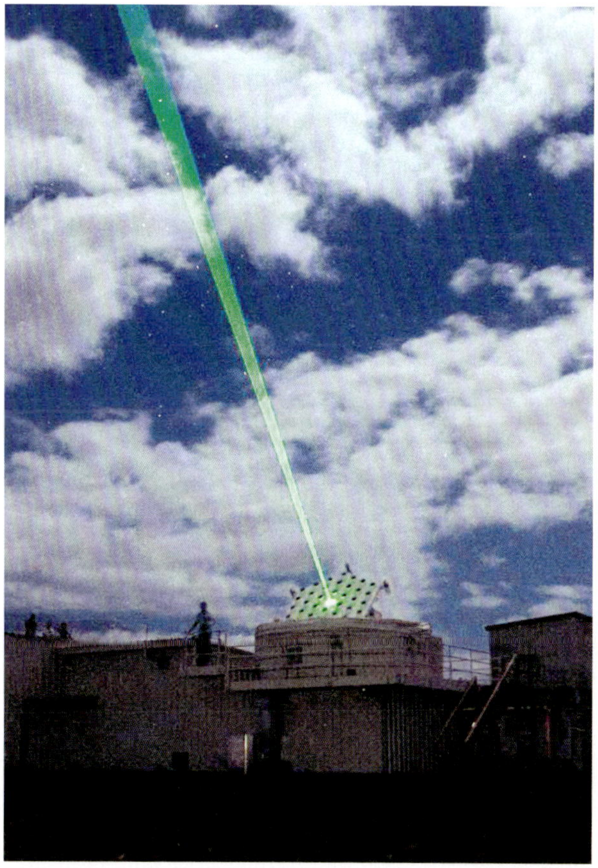

Ball Aerospace played an important role in the U.S. government's Strategic Defense Initiative (SDI) by participating on a team of seven government agencies and private firms in designing, building and operating the relay mirror experiment (RME) in 1990. RME helped validate the technology for high-precision laser pointing and tracking for a ballistic missile defense system. The experiment directed a laser beam from one ground site in Hawaii to another more than 600 miles away by bouncing the laser off a mirror on a satellite orbiting the Earth. The laser beam is pictured here making a direct hit on the target. RME highlighted Ball Aerospace's capability in providing complete aerospace systems and operations.

Eyeglasses for Hubble

Few Ball Aerospace projects for NASA have been more widely publicized or universally praised than its correction of the Hubble Space Telescope's optical system in 1993.

Launched in 1990, Hubble promised to offer new views of the cosmos that would revolutionize scientists' understanding of the origins of the universe. It was expected to give them the ability, as *Design News* explained it, "to see up to 350 times more of the observable universe and analyze events occurring in a period spanning 75% of time since the Big Bang."

In 2003, the Hubble Space Telescope used its Ball Aerospace-made Advanced Camera for Surveys, installed the prior year, to view the Sombrero galaxy 28 million light-years from Earth.

High hopes, however, gave way to despair. Once Hubble was in orbit, scientists discovered a manufacturing flaw in the primary mirror that blurred the telescope's vision. Coming on the heels of four scuttled space-shuttle missions, the $1.5 billion Hubble was criticized as yet another expensive NASA blunder. In the words of *Time* magazine, NASA's ability to fix Hubble "may be the agency's last good chance to prove it can meet a formidable challenge within budget."

NASA turned to Ball Aerospace for help. In a heroic effort, Ball scientist and optics expert Murk Bottema came out of retirement, even though he was terminally ill with cancer, because he had a simple, effective solution: install ten quarter-sized mirrors to refocus and correct the images from the main mirror. Based on that idea, NASA commissioned Ball Aerospace to design and build the Corrective Optics Space Telescope Axial Replacement (COSTAR), known as "eyeglasses for Hubble."

Installing COSTAR on Hubble as the telescope orbited above the Earth presented a special challenge for NASA. Engineer James Crocker of the Space Telescope Science Institute devised an answer while showering at his hotel room during a conference in Germany. The European-style shower head used small arms to adjust the water flow. Crocker realized that similar articulated arms bearing Bottema's mirrors could be extended into Hubble.

Just prior to the launch of COSTAR, Dr. Edward J. Weiler, NASA's chief scientist for Hubble, said, "This project is going into the history books: as a national disgrace or as a great American comeback." It was, in fact, a comeback. After astronauts installed the COSTAR mirrors in a December 1993 spacewalk, Hubble worked perfectly —matching or even exceeding expectations in unveiling the most detailed portraits of the universe ever seen by humankind. In 2004, Congressman Mark Udall said, "Hubble remains one of the most productive scientific instruments in history, and certainly NASA's most productive scientific mission, accounting for 35 percent of all its discoveries in the last 20 years."

Ball Aerospace has continued as a major contractor for Hubble, building a total of seven instruments for it. It is now involved in NASA's next-generation space observatory, the James Webb Space Telescope (JWST). Ball Aerospace has been involved with the JWST program from the beginning as developer of the optical subsystem. Unlike Hubble, which orbits 375 miles above Earth, the Webb telescope will be almost a million miles away. "The Webb telescope may one day help us understand how planetary systems form and interact, and how the universe built up its present chemical/elemental composition," *Mechanical Engineering* magazine noted. "Information from the telescope may also determine the shape of the universe."

Ball Aerospace played a major role in the success of NASA's Hubble Space Telescope, an orbiting astronomical observatory that has revolutionized scientists' understanding of the universe. Since being deployed in 1990, Hubble has been repeatedly upgraded with instruments designed and manufactured by Ball Aerospace. Seven instruments intended for installation on Hubble were Ball-built.

Since 1972 Ball Aerospace has provided the Sea Sparrow low-light camera, used by NATO ships for surveillance and missile fire control in dark or dusk conditions. The Sea Sparrow camera is the small white cylinder mounted atop the twin radar units, right.

environment. Clem Fogelberg, who had left Beech Aircraft Company to join Ball Aerospace in its early days, ran into a former Beech co-worker, John Simpson, at a gas station. Fogelberg said, "You ought to get over to Ball. We're having a lot of fun there." Simpson joined the company weeks later and spent 30 years with Ball Aerospace before retiring in 1994.

The division's initial business was to make pointing controls for rockets being launched from White Sands, New Mexico. Then, in a remarkably quick leap into the space race, Ball received a contract to build NASA's second science satellite, the Orbiting Solar Observatory (OSO), which launched in 1962. Ball won the contract by designing a spinning satellite platform based in part on its rocket pointing controls. The idea was highly innovative, and it earned Bartoe, the project leader, the first Spacecraft Design Award of the American Institute of Aeronautics and Astronautics for "outstanding achievement in the architecture, analysis, design, and implementation of space systems." That satellite became one of seven OSO satellites that Ball built for NASA. The early success of Ball Aerospace "can be traced directly to the pioneering OSO program," Drew Crouch, vice president, strategic development, says.

Ball Aerospace flourished from the 1960s through the 1980s. Its sales were $23 million in 1970 and by 1985 they had reached $160 million. R. H. "Gabe" Gablehouse led the division for much of this period and was largely responsible for broadening its technical capabilities and strengthening its bidding process. Ball Aerospace participated in many exciting and innovative projects, including the Sea Sparrow low-light-level camera designed for North Atlantic Treaty Organization (NATO) ships in 1972 and the Infrared Astronomical Satellite, the first space-based, long-life cryogenically cooled infrared telescope, in 1983.

Well before the launch of Columbia in 1981, Ball began a more than two-decade involvement in NASA's space shuttle program. The first shuttles flew with the aid of Ball's Star Tracker, a navigational system that reads stellar constellations to determine the vehicles' location and position in space.

Since then, Ball has developed additional components for the shuttle itself and its many payloads. These include actuators, mechanical latches that hold shuttle cargo in bays, release it during deployment and receive it again when a mission is complete. Ball also has developed communication antennas used by astronauts during spacewalks, high-resolution imaging systems and a weather-monitoring satellite for NASA's Climatology Program.

However, the early 1990s brought a reversal of fortune that saw Ball Aerospace enter the most difficult period in its history. As communism collapsed and the Cold War ended, the U.S. government trimmed its expenditures for defense and space exploration. To make matters more difficult, some space contracts that once were reserved for American companies began to be awarded to firms in Russia. In 1993, Ball felt the direct impact of this shift when its work on Space Station Alpha was cancelled and the $14 million remaining on its contract was awarded to a Russian company. "Friendship with Russia is not without pain," the *Denver Business Journal* noted wryly in reporting on the cancelled Ball contract.

All of America's premier aerospace companies, not just Ball, struggled in the early 1990s as market conditions changed. From 1990 through 1994, Ball Aerospace trimmed employment by more than 800 people. As results lagged, Ball Corporation hired investment banking firm Lehman Brothers in 1994 to study whether Ball Aerospace should be sold. Lehman contacted potential buyers but could not get a price that Ball was willing to accept. Consequently, Ball decided to keep the aerospace business and improve its performance.

"We couldn't find anyone willing to pay a price we thought was adequate," Hoover says, "and we ultimately decided our aerospace business was worth more to us than to anybody else. It turned out to be a good decision." Ball Aerospace has since emerged as a consistently strong contributor to Ball's profits.

Nonetheless, Wall Street analysts have repeatedly asked Ball Chief Financial Officer Ray Seabrook why the company doesn't sell its aerospace business to concentrate solely on packaging while paying down debt or buying back stock. His kidding reply: Ball should actually sell its packaging business and stick to aerospace, since the latter provides better returns on capital. Owning the two businesses, he says, is like "having two different personalities in your house. Packaging is very structured, neat, tidy and time driven. In aerospace, you may not see a guy for a month and when you ask him what he's been doing he says he's been off designing the intergalactic something or other. What drives a person to work in one would drive a person crazy in the other."

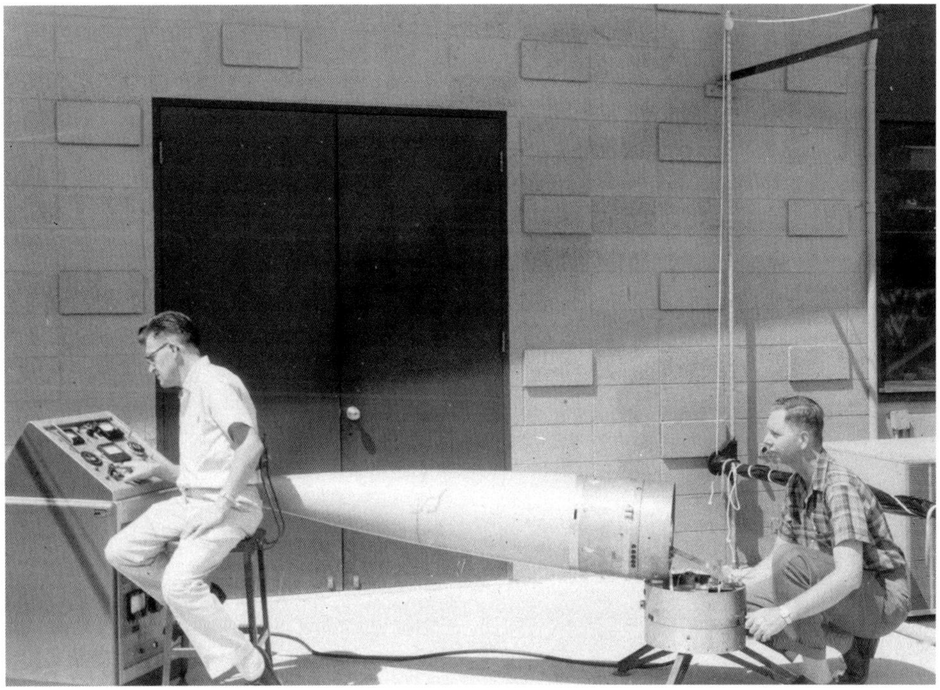

Ball Aerospace introduced its first successful product, a pointing control used by the Air Force to stabilize instruments on scientific rockets and balloons, in 1959. Employees Wayne Burt, left, and Bill Stevens conduct pre-flight tests that year. Drawing on the technical skills developed in this work, Ball Aerospace started winning important NASA contracts in the early 1960s.

Rejuvenating the Business

In late 1994, in a first step to expand revenues, Ball Aerospace created a new subsidiary, EarthWatch, Inc., to serve the newly emerging market for satellite-based remote sensing of Earth. Several months later,

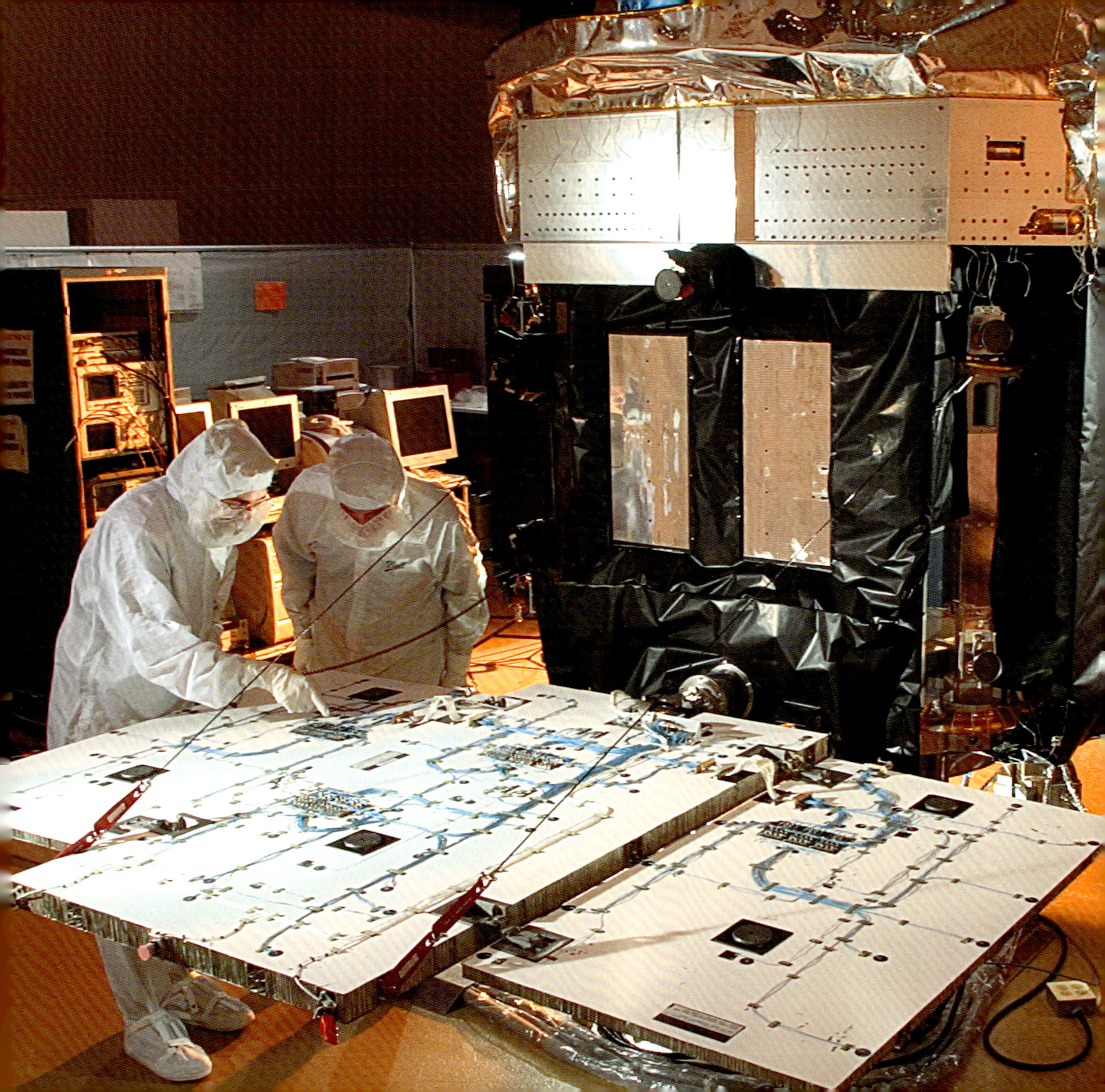

EarthWatch was merged with WorldView Imaging Corporation to form today's DigitalGlobe. Ball holds an equity interest in the company, which provides the highest quality earth imagery available commercially. Ball built DigitalGlobe's first satellite, QuickBird, launched in 2001, and is under contract to build its second, WorldView. DigitalGlobe's detailed photographs of Earth, taken from space, are used by customers for many purposes, such as mapping cities, tracking hurricanes and monitoring the spread of crop disease. Its images dramatically illuminated the devastation of the December 2004 tsunami in Southeast Asia and helped scientists gain valuable insight into the water's movement and its effects upon the land.

Ball Aerospace also streamlined operations to reduce costs, withdrew from the visual image generation systems business and sold its Efratom time and frequency measurement business at a gain. Through these and other actions, including the development of new business opportunities, Ball Aerospace began to recover. "On the ropes in 1993, the subsidiary of Ball Corp. has bootstrapped production through classified government contracts," the *Denver Business Journal* said in 1997. Don Vanlandingham, CEO of Ball Aerospace from 1996 to 2002, built on this success by pushing for growth in commercial markets where technologies developed for the public sector, such as in laser communications, also could be applied.

Ball Aerospace today is focused on three main markets: civil, defense and commercial. "Our core competency [in each market] is innovative design in the application of technology," Dave Taylor says. "We don't do pure research. We meet our clients' needs by applying technology in innovative ways."

Wide Array of High-Tech Projects

Much of Ball Aerospace's defense work is classified. A sampling of projects that have been made public, however, reveals Ball's expertise in imaging, communication and sensing technology along with an extraordinary depth and breadth of capabilities. In fact, some of Ball's projects—especially those for NASA—could have been taken straight out of science fiction.

For example, Ball is building the imaging system and spacecraft, as well as managing all systems operations, for NASA's Kepler Mission, a search for potentially habitable planets. Planned for launch in 2007, Kepler will look specifically for Earth-sized planets likely to contain water by monitoring stars similar to the sun for four years.

In 2004, the Air Force chose a Boeing/Ball Aerospace team to provide mission planning, a satellite and other services for the Space-Based Space Surveillance (SBSS) System. When fully operational, SBSS will consist of a small constellation of satellites that will detect and track all orbiting objects, generating data to support military operations.

Ball Aerospace and NASA's Jet Propulsion Laboratory (JPL) developed the flight hardware and ground systems for NASA's two Deep Impact spacecraft, one of which collided with comet Tempel 1 at the equivalent speed of 23,000 miles an hour on July 4, 2005, spraying some of the solar system's most pristine material, held deep inside the icy comet, out into space. The encounter, taking place 83 million miles from Earth, required such precision that JPL Project Manager Rick Grammier described it as "a bullet trying to hit a second bullet with a third bullet." Deep Impact was the first mission to explore a comet's interior, and scientists expect its spectacular images and analysis will not only provide data on the structure and composition of comets but also shed light on the formation of the solar system.

Ball has played a significant role in all four of NASA's Great Observatories—the Hubble Space Telescope, Chandra X-Ray Observatory, Compton Gamma Ray Observatory and Spitzer Space Telescope—each observing the universe in a different kind of light (visible, gamma ray, X-ray and infrared, respectively). Spitzer, the last of the four, was launched in 2003 to peer deep into space by detecting the infrared energy (heat) radiated by stars,

The Deep Impact instrument platform, above, is inspected by a Ball Aerospace technician. Ball Aerospace built two spacecraft for Deep Impact and also served as the mission's prime contractor. Ball also built NASA's CloudSat spacecraft, whose solar arrays, opposite, are shown during assembly. CloudSat, which launched in 2005, uses advanced radar to measure the altitudes and properties of clouds.

Deep Impact: Smashing Success For Ball Aerospace

Not long ago, the idea of smashing a spacecraft into a comet was considered far-fetched. But on July 4, 2005, after 268 million miles of travel, an impact vehicle the size of a coffee table dropped itself into the path of comet Tempel 1. The resulting collision—at 23,000 miles an hour—dislodged pristine material from below the comet's crust as another spacecraft hovered 300 miles away, collecting images and gathering data. Scientists expect the material ejected, untouched for 4.5 billion years, to provide secrets about the origin of our solar system.

The mission, which the *New York Times* called "a remarkable triumph," was one of the highest-profile projects ever for Ball Aerospace, prime contractor in a collaboration with NASA, the University of Maryland and the California Institute of Technology's Jet Propulsion Laboratory. Ball was the mission prime contractor plus designed and built the "Flyby" and "Impactor" spacecraft and three advanced imaging instruments.

After six years of planning and building, the spacecraft were launched in January 2005. They traveled together until separating 24 hours before reaching the comet. The Impactor continued on its own for 500,000 miles toward the comet while the Flyby moved aside and aligned itself to observe from a distance. The margin for error was small; if the Impactor missed the speeding comet, there would be no second chance. But the mission was flawless, giving scientists their best views ever of a comet and opening a spectacular window into the solar system's past.

Andrew Coates of the Mullard Space Science Laboratory at University College London, principal investigator for two ground-based telescopes observing Deep Impact, called the mission "one of the most audacious experiments that has been undertaken ever." Said Ball Aerospace President Dave Taylor, "We accomplished something thought to be impossible." Detailed analysis of the data obtained by Deep Impact will keep scientists occupied for years.

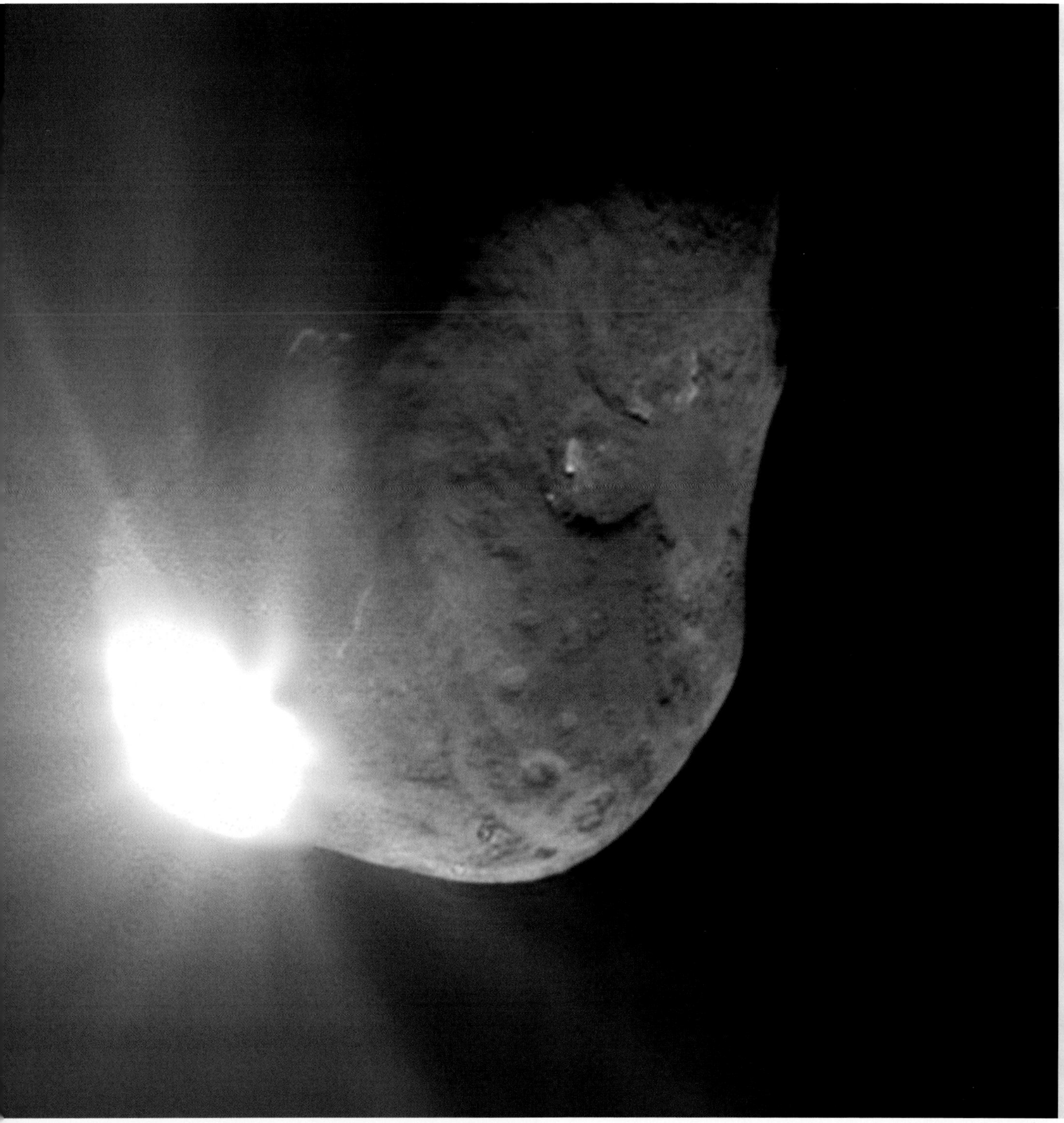

Members of the NASA/Jet Propulsion Laboratory team, opposite, rejoice as Deep Impact's Impactor spacecraft hits a bull's eye on Comet Tempel 1 in July 2005, while its Flyby craft captures a stunning image, left, of the collision. Deep Impact's science team required the Impactor to be built primarily of one of the three noble metals—gold, silver or copper—to ensure that the Flyby craft could get an accurate reading of material ejected from the comet's core. Those three elements were not known to be found on comets and could easily be filtered out of the spectrographic readings used to analyze the material. Thus the Impactor was made of 49 percent copper, the least expensive of the three metals, and 24 percent aluminum—and the mission was commemorated with a copper plaque, opposite.

galaxies and other objects in space. Intergalactic space is filled with vast, dense clouds of gas and dust that cannot be penetrated by optical telescopes. Because infrared light can "see" through these clouds, Spitzer is enabling astronomers to gaze into regions of star formation, the

> "A new focus and a new energy has been put in place at Ball Aerospace in recent years, and that is going to continue."

The color infrared image of Japan's Miyake-jima volcano, opposite, was collected by the QuickBird satellite of DigitalGlobe Inc. Ball Aerospace designed and built QuickBird and is also building the company's next-generation WorldView satellite. DigitalGlobe is a leader in providing high-resolution commercial imagery of Earth as seen from space. Its images are used for mapping, fire risk assessment, environmental studies and many other purposes.

centers of galaxies and newly forming planetary systems. Ball Aerospace built the cryogenic telescope assembly (its helium-cooled "eye") and two key science instruments for Spitzer: the infrared spectrograph, which records the fingerprint of different frequencies of radiation, and the multi-band imaging photometer, a set of arrays that detect far infrared radiation.

Next on NASA's observatory agenda is the James Webb Space Telescope, scheduled for launch in 2013. This "super-telescope" will explore fundamental issues about the earliest epochs of the universe and the processes by which it evolved. Ball Aerospace is developing the advanced optical technology and lightweight mirror system that lie at the heart of the telescope.

Ball Aerospace also is active in NASA's programs to explore Mars. It helped design and build the electronics that control the power aboard the Spirit and Opportunity rovers and also built the camera gimbal, or gyroscopic stabilization system, for the high-gain antenna that relays the rovers' images back to Earth. In addition, Ball developed the high-resolution imaging camera for NASA's High Resolution Imaging Science Experiment (HiRISE) on the Mars Reconnaissance Orbiter, launched in August 2005. The HiRISE mission will investigate

deposits and landforms resulting from geologic and climatic processes and assist in the evaluation of future potential landing sites on Mars.

In October 2004, Ball Aerospace was awarded a NASA contract to build an earth-orbiting observatory as part of a $208 million mission. The Wide-field Infrared Survey Explorer, known as WISE, is scheduled for launch in 2008. It will survey the entire sky in four infrared bands; 1,000 times more sensitive than any previous instrument of its kind, WISE is expected to penetrate interstellar dust to find the brightest galaxies in the universe and the closest stars to our Sun.

Ball Aerospace technology is also monitoring the environment. The company built a payload and spacecraft, respectively, for CALIPSO and CloudSat, scheduled for launch in the second half of 2005. CALIPSO (a joint project of NASA and the French space agency) will study the impact that clouds and aerosols have on the earth's radiation balance. Its data will help create climate predictions that national and international leaders are likely to use to make policy decisions about global climate change. CloudSat (a joint mission of NASA and the Canadian space agency) will measure the altitude and properties of clouds to improve how they are represented in global climate and numerical weather prediction models.

In the defense area, Ball supplies antennas for many military platforms, including Lockheed Martin's F-35 Joint Strike Fighter, the world's most advanced multi-role fighter plane. It also builds cameras that assist military helicopter pilots, manufactures GPS and satellite datalink antennas for the Tomahawk cruise missile and is making antennas for the Marines' next generation of amphibious assault vehicles.

Deep Sense of Pride

Ball's aerospace business has a strong patriotic element. Many employees talk proudly about their company's role in space exploration and national defense and the privileged—even awesome—nature of doing work that includes looking across the universe.

Moreover, employees sometimes have memorable encounters that go beyond designing and making high-tech systems and products. Lisa Pauley, former vice president of finance and administration at Ball Aerospace who is now senior director, administration and compliance for Ball Corporation, tells about the American soldier in Afghanistan in 2002 who needed a high-performance radio antenna made by Ball. Recognizing the advantages of modern wireless communications, the soldier figured the fastest way to get an antenna was by calling Ball Aerospace in Colorado on his cell phone even though he didn't know anyone at the company. The switchboard connected him to an employee who happened to have such an antenna in his office and was happy to send it to the soldier. The antenna was shipped that very day, arriving at the soldier's military post two days later.

Ball Aerospace is a leader in low-observable antennas, above. NASA chose a Ball spacecraft, opposite, for the National Polar-orbiting Operational Environmental Satellite System to study global environmental change.

The Strategic Advantage of Being Small

Ball Aerospace is uniquely positioned in the aerospace industry because of its relatively modest size compared to the industry giants, according to Taylor. "One of our competitors has to win maybe a billion dollars of new business every week not to shrink," he says. "Another has to win maybe half a billion every week. We have to win 12 million." The firm continues to be successful in securing new business and gaining market share.

As it continues to grow, Ball Aerospace is bidding for larger programs. However, as it wins these larger contracts, it faces the challenge of being more rigorous in its business processes and controls without becoming overly bureaucratic and undermining the creative, people-oriented culture that has contributed to its success. Taylor describes that culture as supportive, ethical, caring and honest. "Those aren't clichés either," he insists. "It's okay here to ask questions. It's okay to have issues and problems in your work. If you do, people will help you. I've been in places where you have an issue or problem and people set you adrift mentally and hope you fail. We don't have that. The first order is that people help each other. That's one of the things we bring to the marketplace."

The *Denver Post* agrees. In an article published in October 2003, it said, "Employees say they like the relatively relaxed atmosphere at Ball compared with larger contractors.... Teamwork and collaboration usually supersede internal rivalry and politics." It quoted Rick Ortiz, a hardware designer at Ball Aerospace, as saying, "I've worked other places, and I can say that Ball cares more about its employees."

Ball Aerospace is one of the few remaining aerospace companies that has not been sold to or merged with an industry giant. It continues to go its own way as a successful business within Ball Corporation, contributing substantially to financial results and bringing a sense of pride to employees throughout the company.

"We started before NASA," Taylor says. "There's a long heritage, but we think our best days are still to come."

WITH CHOPPED BEEF
Prime Value
DICED
Tomatoes
SPLAT!
Escape
Go!
Strawberry
LUNA
Jubilee
Water

CHAPTER TEN:

Ball Today: Winning With the Customer

At the heart of Ball's success has always been an uncommon ability to master change, even reinvent itself when necessary. The five Ball brothers demonstrated this early on, growing their company by continually investing in new products, new technologies and new marketing techniques. When Ball stagnated in the 1940s, the next generation of leaders took up the challenge by diversifying into new areas such as plastics and aerospace.

Chairman, President and CEO Dave Hoover says periodic self-examination and reinvention remain as vital to Ball today as ever. In 2002, following the acquisition of Schmalbach-Lubeca, Hoover established the Ball Leadership Team, or BLT, to focus on Ball's future. "I did this," he says, "because I sensed that notwithstanding our success it was time for renewal and reinvention." The group, which meets monthly, brings together 16 corporate officers and business-unit heads from around the world to share ideas, discuss opportunities and challenges and envision the future.

As Ball marked its 125th anniversary in 2005, Hoover and the BLT used the milestone as an opportunity to frame Ball's next steps. "We started with the basic question, although we are 125 years old, why does Ball exist?" Hoover says. The group developed a "core purpose" statement emphasizing that Ball is in business "to add value to all of its stakeholders, whether it is providing quality products and

Ball packaging, used by consumers worldwide every day, safeguards the integrity of beverages and foods while contributing to their effective marketing and efficient distribution. Dave Hoover, left, helped dedicate Ball's expanded packaging research and development facility in Colorado in 2004.

Tim Pratt, opposite, is a millwright at the Saratoga Springs, New York, beverage container plant. He applies his skills to fabricate parts in the plant's machine shop and also works with customers to help ensure that Ball cans run smoothly on their filling lines.

services to customers, an attractive return on investment to shareholders, a meaningful work life for employees or a contribution of time, effort and resources to our communities."

From that purpose emerged a set of core values that have characterized Ball since its founding—and will carry it into the future:

- Integrity
- Respect
- Motivation
- Flexibility
- Innovation
- Teamwork

These values provide a framework for Ball and its people even if the company were to diversify into new businesses beyond packaging and aerospace. They go hand-in-hand with the company's five "keys to success," which offer practical guidance for Ball people in their work:

- *Close to Customers* — "We have a total commitment to being close to our customers and understanding their needs and future direction. This commitment extends throughout our organization."
- *Creativity and Imagination* — "Our employees' creativity and imagination enable us to deliver innovations in products, process development and the way we conduct business so we can better serve our customers, grow the company and increase the value of the enterprise."
- *Behave Like Owners* — "By behaving as true owners of the business, our employees deliver superior results and provide the best value in the products and services we supply to our customers."
- *Attention to Detail* — "By managing our operations with relentless attention to detail we are creating safe workplaces while building a great business that consistently delivers superior value."
- *Build on Strengths* — "We intend to build on our heritage of ethics, integrity, quality and value in all our dealings by treating all stakeholders the way we would like to be treated."

Larry Green, who joined Ball in 1994 to head its plastic bottle business, brings the perspective of someone who has both competed with Ball and worked for it. He says that Ball, even though it is no longer headquartered in Indiana, is "still very Midwestern in its values and is very competitive. We're kind of like Indiana basketball. We don't brag or showboat, but we don't like to lose either. We work hard every day to meet the needs of our customers."

David Westerlund, senior vice president, administration, and corporate secretary, who came on board through the Ball-Incon deal in 1987, says, "Someone wrote that Ball makes decisions slowly. I challenged him on that. I said Ball, in my opinion, makes decisions prudently with thorough thought. Sometimes that means speed, sometimes it doesn't." He describes Ball as being "tenaciously pro-customer."

Greg Dunn, vice president, manufacturing, metal beverage container operations, says, "What really separates us is the relationships we build with our customers. We don't blow our horn about our service, we do it.

"We don't care just about money. I think we have a culture that cares about people and treats them fairly."

More than 80 Ball corporate, packaging and aerospace employees took part in National Rebuilding Day in April 2005, pitching in to improve three homes in the Denver area. Reed Jentzsch of Ball's packaging operations, second from right, and Steve Leitgeb of Ball Aerospace, right, prepare to install a window in a home in Commerce City, Colorado, with the assistance of Frank Keith, center, manager of Linford Glass, which donated the windows. Opposite, a poster created by Ball proclaimed the main message of a recycling campaign in 2005.

When there's a problem, we fix it. We succeed by helping our customers succeed."

Investor Relations Director Ann Scott, who grew up in Indiana and joined the company in 1988, recalls, "When I told my parents, 'I've got a job interview at Ball,' my dad was so excited. He said, 'Now that's a company you can work for the rest of your life.'" She says Ball not only has "very strong corporate values," but also "a wonderful history" on which it builds continuously. The average length of employee service at Ball is 11.6 years, a relatively long average tenure in today's mobile job market. Employee turnover is comparably low, and the company has many long-term employees who have been with Ball for two decades or more.

The company also benefits from its financial strength, evidenced by its strong cash generation and balance sheet. Scott Morrison, vice president and treasurer, points out that Ball has the resources to invest in its existing operations as well as to make acquisitions when they are available on attractive terms and fit the company's strategic needs. He adds, "We feel we've got great access to capital if we need it at very good pricing."

Upholding the Company's Values

Dave Hoover is a champion of the company's values. Like the five Ball brothers, he grew up on a farm. After earning a bachelor's degree from DePauw University and an M.B.A. from Indiana University, he joined Ball in 1970 as assistant to the treasurer. In an age when senior executives often change companies at the drop of a lucrative contract, Hoover is that rare business leader who has spent his entire career with the same company. He became Ball's ninth CEO in January 2001.

Hoover has an outgoing personality and enjoys being with people, and they with him. Moreover, he still has some farm-boy blood running through his veins. He often visits Ball plants and inevitably spends time chatting with employees on the factory floor. Employees at the Findlay, Ohio, plant still talk with admiration and

The Ball Packaging Europe (BPE) Technical Center in Bonn, Germany, develops beverage cans with new functions, shapes and designs. Opposite, BPE's Wilfried Mohr, laboratory technician, packaging technology, left, and Georg Janssen, manager, packaging technology, inspect Ball's sleeker version of the standard 330-milliliter can. The slender silhouette visually embodies a trend toward low-calorie beverages, making the can ideal for "light" and wellness drinks.

glee about the time Hoover visited their plant and noticed that a maintenance worker on one of the can lines had a photograph of his prize hog in his tool box. Hoover stopped and chatted with the worker for several minutes. "They had a healthy conversation about hog farming," according to Larry Coryea, administrative manager at Findlay.

Hoover chuckles when reminded of the incident. His unpretentious manner aside, he's seen as a smart, sophisticated executive with strong leadership skills and personal drive. In 2003, the *Denver Post* described him as "both overseer and architect for a massive expansion that made Ball the biggest beverage and food can maker in the world last year."

Hoover manages the company from a financial perspective, emphasizing that each operation must generate a return on capital that exceeds cost of capital. But he also makes clear, "We don't care just about money. I think we have a culture that cares about people and treats them fairly. However, to be successful and create value for all our stakeholders, we must be profitable and grow. We have more than 13,500 employees who count on us not to mess up, so they can keep working and support their families and send their kids to school."

Ball participates in various markets with growth opportunities, including European beverage containers, plastic bottles and aerospace, as well as in markets that are mature, including North American beverage containers and steel food cans. Through a combination of acquisitions, new products and organic growth, it has more than doubled its revenues and quadrupled earnings in the past five years.

The company seeks to boost earnings per share by 10 percent to 15 percent annually on average over time. "We didn't just make up the 10 percent to 15 percent growth rate," says Ray Seabrook, Ball senior vice president and chief financial officer. About one-third of that projected earnings growth, he says, is expected to come from operational improvements and sales increases and about two-thirds from reinvestment of the company's strong cash flow. "We do not need large revenue increases to meet our earnings targets," he adds. "We need to run our businesses efficiently, maximize our cash flow and invest that cash flow where it will produce the best returns." Even when there are not opportunities to reinvest most of the company's cash flow into the business, it can be used to raise per-share earnings by repurchasing stock and paying down debt.

Ball's growth strategy has found a receptive audience on Wall Street and in the business media. In July 2004, *Fortune* listed Ball as one of its top stock picks. "Glamorous? Hardly," the magazine proclaimed. "But CEO David Hoover runs a highly productive outfit."

In January 2005, *Forbes* noted that Ball used to be "just another small operator in the fragmented commodity packaging business" but is now "a leading global maker of beverage cans" with a growing position in the custom can business. It said Ball shareholders had received a 38 percent annualized return on their investment over the prior five years, the highest among 12 metal products and materials companies surveyed by the magazine.

Emphasis on Innovation

Packaging is one of the world's ubiquitous industries, bringing value to consumers every day by protecting products, ensuring their integrity, facilitating their distribution and providing convenience.

It is not a static business. Major new types of packaging over the past two centuries—including the metal food can in 1813, the home canning jar in 1858, the two-piece beverage can in 1965 and the PET carbonated beverage bottle in 1977—as well as constant innovations in all of these packaging lines, have brought change and created opportunities. Ball did not invent any of these products, but it learned how to manufacture them efficiently. Moreover, the packaging products Ball makes today—beverage cans, food cans and plastic bottles—have enjoyed incredible staying power because they continue

sleek

Ball in the Community

Community involvement is a hallmark of Ball and its employees. Ball people are a generous group, giving back to the community in many ways—locally and globally.

At Ball plants, employees join together to assist local causes. The beverage container plant in Torrance, California, provides volunteer support and donations of materials to a local middle school through an Adopt-A-School program. Employees at other locations volunteer for food banks, help build homes for Habitat for Humanity or participate in charitable walkathons. Employees worldwide contributed to relief efforts following the disastrous Asian tsunami in December 2004. The list goes on and on, reflecting the personal commitment of Ball people to help those in need.

Employees at many Ball facilities have exemplary records of giving to United Way. The DeForest, Wisconsin, food container plant was honored in 2003 for a 517 percent increase in United Way contributions over 2002, the largest gain among local companies. In a campaign covering Boulder, Broomfield and Westminster, Colorado, United Way contributions increased 63 percent after a group of employee volunteers stepped forward to strengthen the fund-raising effort. Employees' charitable contributions in the United States are boosted by the company through matching-gift programs aimed at encouraging involvement in nonprofit and higher education causes.

Employees at Ball's Kent, Washington, beverage can plant collected 640 pounds of aluminum cans in 2003, generating more than $250 that was donated to a local United Way agency. They included, left to right, Joanna Wong, Valerie Jenne, Vicki Buffer (front), Lucy Tacheron, Candy Turner and Brian Thiel.

The company awards college scholarships to children of Ball employees. Six John W. Fisher scholarships are awarded each year in North America, and the program was recently expanded to add five in Europe. With 11 scholarships available in 2005, Ball awarded its 125th scholarship in its quasquicentennial year.

Reflecting its commitment to good corporate citizenship, Ball seeks to operate in an environmentally responsible manner at all its locations worldwide. Many of the company's North American plants—for example, Dandridge, Tennessee; Golden, Colorado; Fairfield, California; Fort Worth, Texas; and Kent, Washington—have consistently received awards for environmental compliance and pollution prevention from state and local organizations. Ball Packaging Europe not only emphasizes environmental compliance, it also actively promotes beverage can recycling. In 2004, it established the RECANS Fund in Serbia. The fund encourages environmental awareness and recycling by supplying teaching materials to schools and organizing workshops and beverage can collection drives.

to serve the needs of product distributors and consumers. None of these packages will disappear any time soon.

The beverage can and other packaging products continue to evolve through research and development. Ball had a new packaging setback in 1994: the Touch Top easy opening can top. The product won the Gold Award for Structural Packaging in an international contest, but was a flop in the marketplace. Ball learned an important lesson: It cannot simply develop new packaging in its laboratories based on its own conceptions. It must work closely with its customers to identify solutions that meet their needs.

R. David Hoover is Ball's chairman, president and chief executive officer. He joined the company as assistant to the treasurer in 1970 and has been CEO since 2001. He has been a champion of the company's values, and helped lead the celebration of Ball's 125th anniversary in 2005.

As a result, partnership with customers is today a core philosophy in Ball's research, just as it is in manufacturing and other areas of the company's business. "To a person, our customers say Ball takes partnership to a new level," John Hayes says. "We talk about 'win with the winners.' Another way of saying that is, we win when they win."

In 2004, Ball combined all its North American packaging research and development in an expanded modern, centralized facility in Westminster, Colorado, to seek improvements in existing packages and explore opportunities in new types of packaging. The *Rocky Mountain News* said Ball put "all the brains and machines it relies on to dream up new cans and plastic bottles" in one centralized facility under one roof. The newly expanded research facility highlights Ball's determination "not only to be a great manufacturer but to move a step beyond that and become more innovative in the development of new products," says John Friedery, senior vice president of Ball and chief operating officer of its North American packaging operations. Called the Edmund F. Ball Technology & Innovation Center in honor of the late CEO Ed Ball, the R&D laboratory is able to replicate the stages of the manufacturing process and test new types of metal and plastic containers, as well as develop better ways to produce them.

Efforts at the research center in Colorado are coordinated with those at Ball Packaging Europe's technical center in Bonn, Germany. Major areas of focus at Bonn include the development of "active" beverage cans that communicate with the consumer; optical effects that can positively influence consumers' purchasing decisions at the point of sale; and innovations that offer functional value, such as reclosable cans.

Positioned for Success

Ball has come a long way in the past 125 years, growing dramatically in some periods and persisting through difficulties in others—but always focused on the needs of its customers. It continues to set a standard of excellence in the packaging industry.

Friedery says, "We're a largely unknown company. Low key. Stealthy. Not very sexy. People don't know much about us. But we are very successful."

Hoover points out that Ball has a mix of experienced employees and newer blood. About half its 13,500 employees are new to the company since 1998, including those that joined Ball in the Reynolds and Schmalbach-Lubeca transactions and new hires in the aerospace and other businesses. He believes this combination of experience and fresh ideas will continue to serve the company well.

He adds, "We have a wonderful company with talented people and a great track record. So long as our people are focused on the same goals, we can accomplish anything."

Directors

(left to right, front row)

George M. Smart
Retired president of Sonoco-Phoenix of Canton, Ohio

Jan Nicholson
President of The Grable Foundation of Pittsburgh

Hanno C. Fiedler
Executive vice president of Ball Corporation; chairman and chief executive officer of Ball Packaging Europe

Howard M. Dean
Retired chairman of the board of Dean Foods Company of Dallas

(left to right, back row)

John F. Lehman
Chairman of J.F. Lehman & Company of New York City

George A. Sissel
Retired chairman of the board of Ball Corporation

Erik H. van der Kaay
Retired chairman of the board of Symmetricom of San Jose, California

Stuart A. Taylor II
Chief executive officer of The Taylor Group L.L.C. of Chicago

R. David Hoover
Chairman of the board, president and chief executive officer of Ball Corporation

Theodore M. Solso
Chairman and chief executive officer of Cummins Inc. of Columbus, Indiana

Company Officers

Charles E. Baker
Vice president, general counsel and assistant corporate secretary

Douglas K. Bradford
Vice president and controller

Hanno C. Fiedler
Executive vice president, Ball Corporation; chairman and chief executive officer, Ball Packaging Europe

John R. Friedery
Senior vice president, Ball Corporation; chief operating officer, North American packaging

John A. Hayes
Executive vice president, Ball Packaging Europe

R. David Hoover
Chairman of the board, president and chief executive officer

Scott C. Morrison
Vice president and treasurer

Raymond J. Seabrook
Senior vice president and chief financial officer

Harold L. Sohn
Vice president, corporate relations

David A. Westerlund
Senior vice president, administration, and corporate secretary

Operations Executives

Brian M. Cardno
President, metal food container operations

Jan Driessens
President, Ball Packaging Europe

Larry J. Green
President, plastic container operations

Michael D. Herdman
President, metal beverage container operations

David L. Taylor
President and chief executive officer, Ball Aerospace & Technologies Corp.

Terence P. Voce
Chairman and chief executive officer, Ball Asia Pacific Ltd.

Directors and Officers: 1980-2005

Aschenbrenner, Frank A. (2)

Baker, Charles E. (2)

Ball, Edmund F. (3)

Bell, R. Donald (2)

Bell, Thomas D., Jr. (1,2)

Bracken, Alexander M. (3)

Bracken, Frank A. (1,2)

Bradford, Douglas K. (2)

Campbell, Roland P. (2)

Clark, Thomas B. (2)

Davis, Delmont A. (1,2)

Dean, Howard M. (1)

DiLoreto, David A. (2)

Durbin, Richard E. (2)

Emerson, Duane E. (2)

Fiedler, Hanno C. (1,2)

Fisher, John W. (1,2,3)

Friedery, John R. (2)

Gaiser, R. Arthur (1)

Gillam, Larry T. (2)

Gillett, Richard M. (1)

Goodrich, Henry C. (1)

Grandcolas, Leon M. (2)

Haas, John A. (2)

Hackett, John T. (1)

Hannah, William C. (2)

Hansen, Arthur G. (1)

Hayes, John A. (2)

Hicks, Donovan B. (2)

Hinsell, Lars (2)

Hoover, R. David (1,2)

Lehman, John F. (1)

Lewis, Donald C. (2)

Lincoln, William A. (2)

Looney, H. Ray (2)

McFadden, Betty M. (1)

McFadden, George (1)

McVie, A. Malcolm (1)

Matsik, George A. (2)

Mercure, Ruel C., Jr. (1)

Midgett, Leon A. (2)

Miller, Barbara J. (2)

Mohlman, Robert H. (1,2)

Morrison, Scott C. (2)

Nicholson, Jan (1)

Overmyer, Elizabeth A. (2)

Owsley, Jr., Alvin (1)

Peterson, William L. (1,2)

Poling, Douglas E. (2)

Pruis, John J (2)

Ringoen, Richard M. (1,2)

Schlesinger, Albert R. (2)

Seabrook, Raymond J. (2)

Sheldon, David B. (2)

Sissel, George A. (1,2)

Smart, George M. (1)

Snyder, John H. (2)

Sohn, Harold L. (2)

Solso, Theodore M. (1)

Spire, Robert M. (1)

Staley, Delbert C. (1)

Stephens, W. Thomas (1)

Stiritz, William P. (1)

Taylor, Stuart A., II (2)

van der Kaay, Erik H. (1)

Voran, Reed D. (1)

Westerlund, David A. (2)

Wild, Charles E. (2)

(1) Director
(2) Officer
(3) Director Emeritus

Acknowledgments

Writing a corporate history is a collaborative process. Just as many people contribute to the success of a company, the story of that success can only be told with their help.

I interviewed more than 100 current and retired Ball employees for this book. I found among them a deep sense of pride in the values that Ball stands for, including consistent product quality and ethical operations. Without exception, they were extremely gracious in sharing their time and ideas and helping me learn about the company. I visited several Ball manufacturing facilities. In every case, the plant managers and employees made me feel like a welcome guest. I am deeply grateful to all who helped, and wish, in particular, to thank the following:

CEO Dave Hoover gave graciously of his time in answering my questions and discussing the company's operations, strategy and outlook. Harold Sohn, vice president, corporate relations, was a strong advocate for the book and worked with me to identify key subject areas and events. Scott McCarty, director of corporate relations and the manager of this project, was a writer's dream—helping me understand the company, guiding development of the manuscript, and setting me up with all the right people and information I needed to do my job.

I have very much enjoyed working with the people of Ball and hope this book captures the qualities that set the company apart as a leader in packaging and aerospace.

— *Richard Blodgett*

New York

October 2005

Timeline

1880
Ball Corporation begins as Wooden Jacket Can Company in Buffalo, New York

1884
Company makes its first home canning jars

1886
Company is incorporated as Ball Brothers Glass Manufacturing Company, owned by the five Ball brothers: Edmund, Frank, George, Lucius and William

1886
Glassmaking plant opened in Muncie, Indiana, which becomes corporate headquarters

1892
As sales of canning jars grow, Ball's workforce exceeds 1,000

1898
F.C. Ball machine, the first semi-automatic glassmaking machine, introduced

1900
Ball is now the world's largest canning jar manufacturer

1909
Company publishes its first *Ball Blue Book*, featuring home-canning instructions and recipes

1912
Company opens zinc mill in Muncie to manufacture canning-jar tops

1914
Company begins making rubber sealing rings for jars

1916
Following the acquisition of a paper mill in Muncie, company makes corrugated cartons in which to ship its canning jars

1917
The Ball brothers purchase the campus of a defunct college in Muncie and turn it over to the state which creates, with Ball family support, today's Ball State University

1921
Co-founder William C. Ball dies

1922
Name changed to Ball Brothers Company

1925
Co-founder Edmund B. Ball dies

1926
Diversification continues with the manufacture of "B" and "C" battery cans

1931
Sales of Ball canning jars peak at nearly 190 million, equivalent to 1.6 jars for every man, woman and child in the United States

1932
Canning jar sales plummet as the Depression takes hold

Co-founder Lucius L. Ball dies

1933
To diversify, company begins making food jars and beer and liquor bottles

Ball family forestalls a potential run on Muncie's banks by pledging its financial support to keep the banks solvent

1934
Muncie zinc mill expanded, becoming largest strip-rolling mill in world; zinc sales to outside customers are increased

1935
Rubber department expanded to make components for household appliance and automotive industries

1942-1945
Company produces ammunition shells and more than half the zinc battery casings used by American armed forces during the war

1943
Co-founder Frank C. Ball dies and is succeeded as company president by his brother George

1946
Ball Brothers Company struggles financially in the postwar years; it begins to modernize facilities and diversify into new products

1948
Edmund F. Ball, son of co-founder Edmund B. Ball, becomes CEO

1955
Co-founder George A. Ball dies

1956
Ball Brothers Research Corp. subsidiary (later renamed Ball Aerospace & Technologies Corp.) is formed in Boulder, Colorado, to participate in technology fields

1960
Ball Brothers Research is selected to build an Orbiting Solar Observatory satellite, its first of many NASA contracts

1961
Glass plant opened in Asheville, North Carolina

1962
Plastic container division established

Muncie glass plant closed

1969
Ball enters the beverage can business – today its largest business – with the acquisition of Jeffco Manufacturing Company of Golden, Colorado

Company name changed to Ball Corporation

1970
John W. Fisher becomes CEO

Large, modern zinc mill completed in Greeneville, Tennessee

1971
Company expands its beverage can operations by building a plant in Williamsburg, Virginia

1972
Company goes public with $26 million offering of stock

Beverage can plant built in Findlay, Ohio

Ball Aerospace designs and manufactures Seasparrow low-light-level cameras used by NATO ships for surveillance and missile fire control

1973
Ball shares listed on New York Stock Exchange under ticker symbol BLL

1975
Agricultural systems division formed

1976
Beverage can plant built in Fairfield, California

1978
Unimark Plastics acquired

1981
Richard M. Ringoen becomes CEO

First Space Shuttle is launched, featuring Ball star trackers, antenna and actuators

Beverage can plant built in Saratoga Springs, New York

Company begins producing zinc penny blanks for U.S. Mint

1984
Ball Aerospace supplies the spacecraft bus and Stratospheric Aerosol Gas Experiment (SAGE II) instrument for the Earth Radiation Budget Satellite (ERBS); expected to have a two-year life, ERBS is still providing – more than 20 years later – valuable data about the Earth's ozone layer and the heat balance of the planet.

Sales reach $1 billion

1986
Company begins producing beverage cans in People's Republic of China through a joint venture in Guangzhou

1987
Beverage can plant built in Conroe, Texas

Company merges glass container operations with those of InCon Packaging to form Ball-InCon Packaging Corporation, a 50-50 joint venture

1988
Company enters metal food can business by forming Ball Packaging Products Canada joint venture

1989
Joint venture formed to manufacture beverage cans in Taiwan

1990
Company acquires partner's interest in Ball-InCon Packaging, becoming sole owner

1991
Company acquires full ownership of Ball Packaging Products Canada

Delmont A. Davis becomes CEO

Sales exceed $2 billion

1992

Company adopts EVA® (Economic Value Added) management discipline, aimed at creating value for shareholders

Commercial glass operations of Kerr Group, Inc., acquired

1993

Seven product lines, including home canning jars, divested in Alltrista Corp. spin-off

Company expands its food can business by acquiring Heekin Can, Inc.

Ball Aerospace designs and builds the corrective technology that restores the imaging of the Hubble Space Telescope

1994

Ball enters the PET plastic bottle business, announcing plans to build factories

Joint venture formed to manufacture beverage cans in the Philippines

George A. Sissel becomes CEO

1995

Joint venture formed to manufacture beverage cans in Brazil

Company exits commercial glass business

Ball's first PET plastic bottle plant opens in Chino, California

EarthWatch joint venture established to provide high-resolution satellite imaging of the Earth

1996

PET plastic bottle plants opened in Ames, Iowa, and Baldwinsville, New York

Joint venture formed to manufacture beverage cans in Thailand

1997

Company becomes largest beverage can manufacturer in China by acquiring M.C. Packaging Ltd.

PET plastic bottle plant opened in Delran, New Jersey

Two bottle-making plants purchased from Honickman Group

1998

In its largest corporate transaction to date, Ball acquires Reynolds Metals Company's beverage can operations, with 16 plants in 12 states and Puerto Rico

Corporate headquarters relocated from Muncie, Indiana, to Broomfield, Colorado

1999

Ball and ConAgra Grocery Products Company form joint venture, Ball Western Can Company, to supply metal food cans for ConAgra Grocery Products

Sales exceed $3 billion

2001

R. David Hoover becomes CEO

Ball and Coors Brewing Company form joint venture to supply approximately 4.5 billion aluminum beverage cans and ends per year to Coors

PET bottle maker Wis-Pak Plastics, Inc., acquired

2002

Company expands globally by acquiring Schmalbach-Lubeca AG, Europe's second largest beverage can maker

2003

Ball Aerospace provides the high-gain antennas and gimbals, pancam mast assemblies, and power and avionics electronics for the Mars Exploration Rover Missions

NASA's Spitzer Space Telescope, using a cryogenically cooled telescope built by Ball Aerospace, sees through thick space dust to return unprecedented views of the universe

Metal Packaging International beverage can ends plant in Northglenn, Colorado, acquired

Sales exceed $4 billion

2004

Company strengthens its engineering, technology and product development skills by expanding the Edmund F. Ball Technology and Innovation Center in Westminster, Colorado

Company acquires full ownership of Ball Western Can Company with continued long-term agreement to supply metal food cans to ConAgra Foods

Sales exceed $5 billion; earnings reach a new high of $296 million

2005

The Deep Impact spacecraft pair, designed and built by Ball Aerospace for NASA, intercept Comet Tempel 1 and excavate material, providing the first-ever look at the primordial nucleus of a comet

Ball celebrates its 125th anniversary

Index

Bold listings indicate illustrated material.

A

B

C

D